GUIDE TO THE GEOLOGY

OF

COLORADO

BY

ANDREW M. TAYLOR

Published by Cataract Lode Mining Company
2108 Arapahoe Street
Golden, CO 80401

First Edition August, 1999

For Ordering Information (Wholesale Only), Contact:

Cataract Lode Mining Company
2108 Arapahoe Street
Golden, CO 80401
303/279-3053

ISBN 0-9634184-5-9

COVER PHOTO: Pike's Peak as viewed from the Garden of the Gods.

TABLE OF CONTENTS

BOXES

FIGURES

TABLES

INTRODUCTION

For practical (and historical) purposes, serious geological study of Colorado began with the 1858 discovery of gold along the Platte River near Cherry Creek (the future site of Denver). This discovery started the famous "Pikes Peak or Bust" gold rush complete with prospectors and other "Fortune Hunting Types" who were everpresent and everready to relieve the prospectors of their hard-earned gains.

This time period in Colorado history (last half of the Nineteenth Century lasting into the Twentieth Century) left us with an abundance of modern-day reminders of the former activities, such as the following:

> Legendary tales of fortunes and hardships

> Legendary mining areas and towns such as Central City, Blackhawk, Gold Hill, Leadville and Cripple Creek.

> Legendary figures such as the Silver Kings: Horace and Baby Doe Tabor, Meyer Guggenheim and J.J. Brown (husband of the "Unsinkable" Molly Brown of Titanic fame), and others.

Along with, and as a result of, gold mining activities came the discovery of silver, copper, lead, zinc, molybdenum and uranium. These discoveries added "fuel" to the gold-fever atmosphere that prevailed during this period of Colorado history.

Some of the earliest, organized geologic activities in Colorado began with the 1873 Hayden Survey. This survey was composed of scientists representing geology, paleontology, mineralogy, botany and zoology, and included topographers, an artist and the famous photographer W.H. Jackson. In addition to the technical expertise necessary for each discipline represented, hardy individualism was a requirement as well as the ability to shoot a gun and the skill to ride a horse or a mule (See Figure 1).

Scientific field work in the latter half of the Nineteenth Century required that individuals endure extreme hardships and extreme conditions as well as ever present dangers such as wild animals, inclement weather, frontier medicine and often unfriendly Native Americans. By the turn of the century, the need for guns had diminished somewhat, but the need for horses and mules had not (See Figure 2). After World War II, more roads were built and 4-wheel drive vehicles began to appear. These factors, augmented by modern prospecting advancements and other "modern conveniences" heralded the rapid decline of the "old time" prospector and the ambience associated with these hardy pioneers.

This guidebook was conceived and written with the beginning *geology* student and the geology enthusiast in mind. Simplification, without oversimplification of the exceedingly complex Colorado geology has been attempted. The guidebook was designed to be useful, interesting and relatively painless for readers and students, but was also designed to include enough geologic material to make scientific sense of the topics discussed. Readers should understand that every *rock* has a story to tell, some mundane and some exceedingly exciting. It is the task of the geologists to unearth these stories using available tools; this guidebook is one of those tools.

This land we call Colorado has changed dramatically throughout time. It has been covered repeatedly by oceans, has experienced climatic changes ranging from tropical/temperate to alpine/glacial. The land has been blasted by explosive *volcanoes* which covered it with *ash* and *lava*. Colorado has suffered innumerable large *earthquakes* as the mountains slowly pushed their way up over millions of years. Mountain ranges have risen only to be slowly eroded away. A giant fireball from a meteorite impact even burned the area 65 million years ago. All-in-all, one could say that Colorado has had its good days as well as its bad days!

When visitors to the scenic areas of Colorado see something that piques their interest, they ask very obvious and profound questions. What is that? What is it made of? How did it form? The answers to some of .those questions are found in this book.

This publication contains a glossary, various maps and diagrams, a geologic time scale and a stratigraphic correlation chart, all of which are intended to aid the reader. Definitions of italicized terms and concepts are presented in the glossary. The Geologic Time Scale (Figure 3) gives the subdivisions of time. Geologic and geographic features are illustrated on maps contained in the pocket (Figure 4, Geologic Map of Colorado and Figure 5, Location Map). Diagrams present geologic models. *Formation* names used across the state are shown on the Stratigraphic Correlation Chart (Figure 6).

This guidebook has been designed to be studied in its entirety, for those interested in the complete story. or to be used as a study and/or field guide for individual topics or areas of interest. Whichever fits your needs or interest, enjoy yourself learning the information presented.

Figure 1. Geologists in Colorado. An early US Geological Survey field party in Colorado. Clearly a field geologist in those days had to shoot a gun and ride a mule! The gentlemen in the photo are identified as Endlich, Rhoda, and Wilson. Photo by Jackson, W.H., 1874. USGS Photo Library.

Figure 2. Moving Camp. Mule-drawn covered wagon at the top of Berthoud Pass. The mules earned their oats that day! Early USGS geological field parties stayed in tent camps. When they moved to a new area, their camps were moved by covered wagons. Berthoud Pass is located on U.S. Highway 40 south of the Winter Park resort area. Thousands of tourists and skiers travel this route today which once was only accessible by horses and wagons. Photo by M.R. Campbell. No date. USGS Photo Library.

ABOUT THE AUTHOR

Andrew M. Taylor, Ph.D., is a graduate of Colorado School of Mines. Since 1989 he has taught *geology* at Metropolitan State College of Denver. His classes are known for their wide ranging field trips throughout the state. One of his most popular courses is entitled (guess what!) "Geology of Colorado". The average student taking this course is a non-science major in need of general studies credit in science. Over the years, Dr. Taylor has been faced with the challenge of presenting complex geology to non-geologists. This publication is a direct outgrowth of these teaching efforts.

Dr. Taylor engaged in mineral collecting in the Tarryall Mountains in June of 1996. The Tarryall Mountains are located west of Colorado Springs. Photo courtesy of Jeff Borth.

ACKNOWLEDGEMENTS

I would like to express appreciation for all of the valuable assistance provided in the creation of this publication. I will always be indebted to John Kilcoyne and Duane Moredock for their excellent advice and guidance. Excellent pencil drawings of fossils were provided by Clint Lemmons. These drawings add much to the text. I am extremely grateful to Glenn Andersen, Bill Chirnside and Maurice Pattengill for critical review of the manuscript. Editing a manuscript is a laborious and seemingly thankless task which is critical to the success of a publication. These gentlemen may lay claim to many free lunches as a reward for their excellent service!

The Geologic Map and the Location Map were prepared using GIS on the computer. I received excellent instructions into the mysteries of ArcInfo and Arcview from Professor Rafael Moreno of Metropolitan State College of Denver and a number of his students, including Anita Hoover, Kit Dwyer, Gregg Rossi and Douglas Steinshower.

The author is also indebted to the many students who have "product tested" the material presented in this book.

FIGURE 3.   GEOLOGIC TIME SCALE

ERA	PERIOD			EVOLUTION OF LIFE		TIME BEFORE PRESENT IN MILLIONS OF YEARS
CENOZOIC	QUATERNARY			Ice ages and humans.		2
	TERTIARY	Pliocene		Development of mammals.		5
		Miocene				24
		Oligocene				38
		Eocene				55
		Paleocene				65
MESOZOIC	CRETACEOUS			Extinction of dinosaurs. Development of flowering plants.		138
	JURASSIC			Climax of dinosaurs. First birds.		205
	TRIASSIC			First dinosaurs and small mammals.		240
PALEOZOIC	PERMIAN			Abundant conifers. Reptiles developed.		290
	PENNSYLVANIAN			First reptiles. Abundant insects. Great coal-forming swamps.		330
	MISSISSIPPIAN			Abundant crinoids.		360
	DEVONIAN			First amphibians. Earliest forests.		410
	SILURIAN			First land plants.		435
	ORDOVICIAN			Primitive fishes.		500
	CAMBRIAN			Corals, brachiopods, trilobites.		570
	PRECAMBRIAN TIME			First primitive single-celled organisms........		3,500
				Origin of Earth................		4,600

Age Dates Conform To Those Currently Used By the U.S.Geol.Survey

SIMPLIFIED COLORADO STRATIGRAPHIC CORRELATION CHART

ANDREW M. TAYLOR

ERA	PERIOD		WESTERN COLORADO		CENTRAL COLORADO	EASTERN COLORADO
			SOUTHWESTERN	NORTHWESTERN		
CENOZOIC	TERTIARY	PLIOCENE		BROWNS PARK FM	THIRTY-NINE MILE VOLCANICS and OTHER ROCKS	OGALLALA FM
		MIOCENE	OTHER NAMES			WHITE RIVER FM
		OLIGOCENE				CASTLE ROCK FM / FLORISSANT LAKE BEDS
		EOCENE		UINTA FM		DAWSON FM / GREEN MTN FM
			SAN JOSE FM	GREEN RIVER FM		
		PALEOCENE		WASATCH FM	DENVER FM	
			ANIMAS FM	FT. UNION FM		DENVER / ARAPAHOE FM
MESOZOIC	UPPER CRETACEOUS		KIRTLAND SH	MESAVERDE GP		LARAMIE FM
			FRUITLAND FM			FOX HILLS FM
			PICTURED CLIFFS SS			PIERRE SH
			MESAVERDE GP		PIERRE SH	
					NIOBRARA FM	NIOBRARA FM
			MANCOS SH	MANCOS SH	COLORADO GP	COLORADO GP or BENTON SH
			DAKOTA SS	MOWRY SH	MOWRY SH	MOWRY SH
	LOWER CRETACEOUS			DAKOTA GP	DAKOTA GP	DAKOTA GP / SO. PLATTE FM / LYTLE FM
			BURRO CAN.-CEDAR MTN.			
	JURASSIC		MORRISON FM	MORRISON FM	MORRISON FM	MORRISON FM
			SAN RAFAEL GP and OTHER NAMES	SAN RAFAEL GP		RALSTON CK FM
			ENTRADA SS	ENTRADA SS		
			CARMEL FM			
		GLEN CANYON GP	NAVAJO SS	GLEN CANYON GP		
			KAYENTA FM			
			WINGATE SS			
	TRIASSIC	DOLORES	CHINLE FM	CHINLE FM		
			SHINARUMP FM	SHINARUMP FM	SHINARUMP FM	LYKINS FM
			MOENKOPI FM	MOENKOPI FM	MOENKOPI FM	

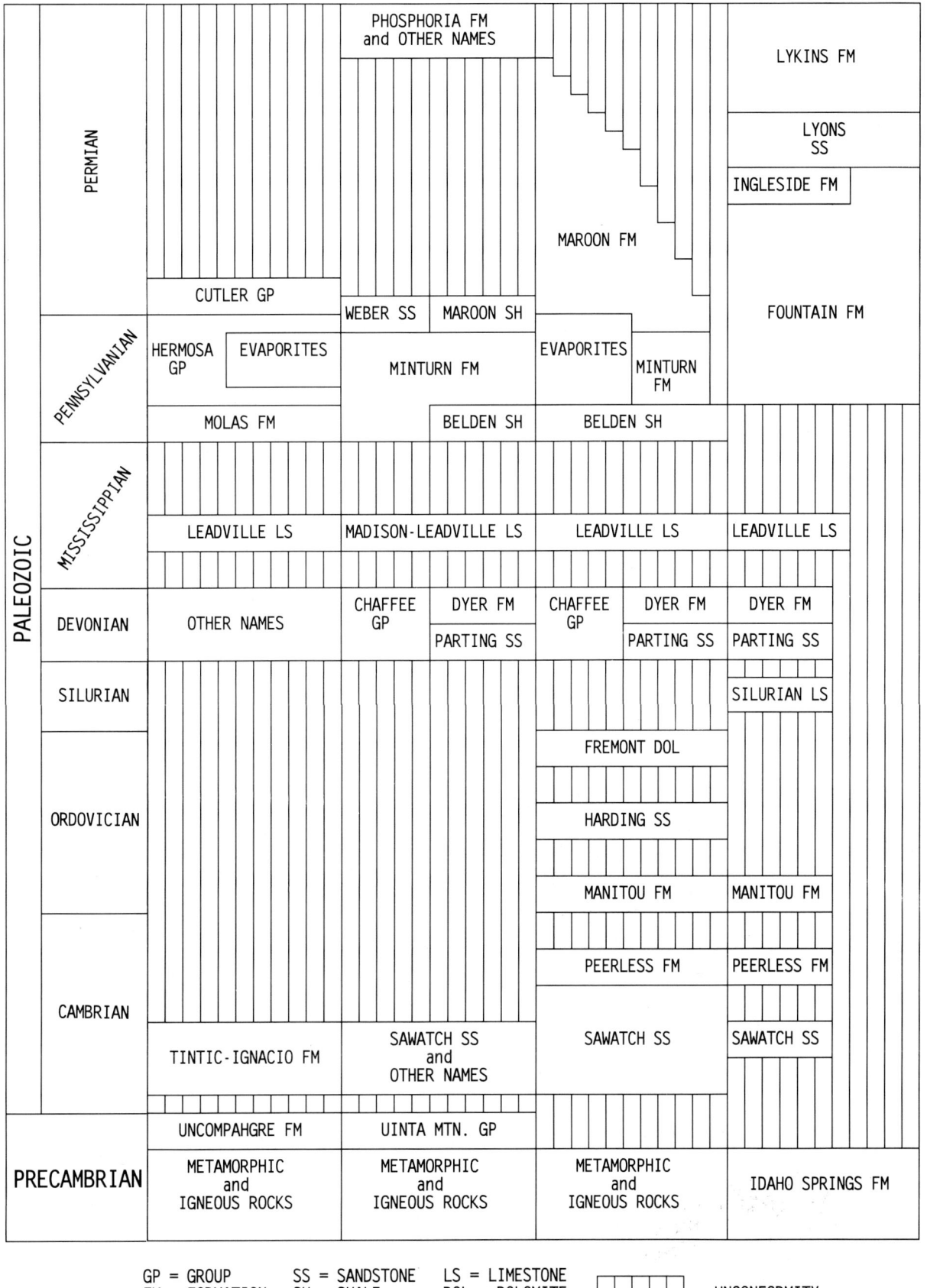

GP = GROUP   SS = SANDSTONE   LS = LIMESTONE
FM = FORMATION   SH = SHALE   DOL = DOLOMITE   = UNCONFORMITY

Figure 6. Simplified Colorado Stratigraphic Correlation Chart. The stratigraphic correlation across the state of Colorado is extremely complex. There are many named formations and rock units. The author has greatly simplified the stratigraphy presented in this chart to avoid overwhelming the casual reader. The named rock units were chosen based upon (1) regional extent and (2) importance. These characteristics are, of course, a subjective judgement. The author apologizes if he has omitted anyone's favorite formation! Not all of the named rock units presented are discussed in the text. Sources of data include numerous publications by the Rocky Mountain Association of Geologists.

The gentleman in the photograph is Harry Yount, the first ranger of Yellowstone National Park. The view is to the north from above Berthoud Pass. Harry apparently visited the Hayden Survey in Colorado in 1874. The picture was taken by W.H. Jackson. USGS Photo Library.

GEOLOGIC PROVINCES OF COLORADO

AN OVERVIEW

Colorado is composed of the following three geologically controlled geographic provinces:

1. Mountains: Central portion of state.
2. Plateaus: Western portion of state.
3. Plains (prairies): Eastern two/fifths of state.

The mountain province has largely been created by strong vertical uplifts accompanied by emplacement of *intrusive* and *extrusive igneous rocks*. It was later modified by *glaciation* and normal processes of *weathering* and *erosion*. The *plateaus* were uplifted vertically and then dissected by stream erosion. The plains have been a site of deposition of sediments throughout most of Paleozoic time. These sediments came from erosion of various highland areas and from deposition of sediments in the many oceans which have covered the land over time. The plains have not been deformed by structural movements of the earth's *crust*.

MOUNTAINS:

The mountains of Colorado occupy, more or less, the central portion of the state. On the east side, mountains rise abruptly along a north-south line. The western border is a much more irregular line which generally follows a north-south trend. Colorado boasts numerous mountain ranges which contain peaks that rise majestically above timberline. Fifty-three of these high peaks reach elevations of 14,000 plus feet. The highest of these is Mt. Elbert with an elevation of 14,443 ft. The spectacular continental divide follows the summit ridges of many of these higher peaks. East of the divide, water drains to the Atlantic. West of the divide, water drains to the Pacific. The headwaters of Colorado's major rivers have their origin in the high mountains. The South Platte and Arkansas rivers flow to the east and the Rio Grande River flows to the south. The mighty Colorado River and the Yampa flow to the west. The climate of the mountains is generally humid, cool in the summers and downright cold in the winters. Much of the precipitation falls as snow in the winter. During December and January, some areas have essentially an arctic climate. Visitors should understand that where trees naturally exist in Colorado, it snows. The more trees, the more snow! Areas without trees below timberline experience less snow.

PLATEAUS:

The *plateau* country of Colorado lies west of the mountains.
It is a relatively flat country with table top *mesas* and deep river
eroded canyons. Elevations are generally 5,000 feet above sea
level and higher. Some mesas have elevations as high as 11,000
feet. The climate is arid and is on the average warmer than the
mountains.

PLAINS:

The sprawling plains east of the mountains have a gently
rolling or hilly appearance and range from 3,350 to 5,000 feet in
elevation. Two major rivers drain eastward, the South Platte and
the Arkansas. The climate is arid with 15 to 16 inches per year of
precipitation. A desert is defined as an area with less than 10
inches of moisture per year. The plains are not a true desert but
they are close.

The famous dinosaur trackway at Dinosaur Ridge near
Morrison. The larger, broader tracks are thought to have
been made by Iguanodon, a herbivore. The smaller tracks
with narrow toes were made by carnivorous theropods of
unknown species.

GEOLOGIC HISTORY

The earth is believed to be 4.6 billion years (*B.Y.*) old which is the age of the Solar System. Meteorites which have come to earth from within the Solar System are of this age. The oldest rocks found on earth are 4.3 plus B.Y. old. Rocks known to exist in Colorado range from approximately 2.1 B.Y. old up to very recent with many gaps (*unconformities*) in the record.

The rocks of Colorado are composed of three major types: *igneous*, *metamorphic* and *sedimentary*. Igneous rocks are formed by cooling of *magma* which has come from deep within the earth. *Intrusive* igneous rocks are those which crystallized from magma underground at depths usually greater than 5 kilometers. *Granite* is a typical intrusive igneous rock. *Extrusive* igneous rocks (also called *volcanics*) are *lavas*, *ashes* and *tuffs* which poured or blew out of *volcanoes*. Metamorphic rocks are those which have been altered to distinct new rocks by heat and pressure. In the process of *metamorphism*, old *minerals* in the rock decompose and new ones are created. The new rock is texturally different from the original (the *parent rock*) but the chemistry is the same. With regard to the age of parent rocks, all that can be said is they are older than their offspring (the metamorphic rock). The most common metamorphic rocks in Colorado are *gneiss* and *schist*.

Igneous and many metamorphic rocks are created by what geologists call *tectonic* activity (the movement of rocks and magma in the earth's crust). The forces and mechanisms of tectonic activity are explained in the *Theory of Plate Tectonics*. For more detailed information on Plate Tectonics, see Box 1.

BOX 1. PLATE TECTONICS

To the people living on the face of the earth, it seems to be a very stable place. Continents, mountains and oceans do not appear to move. But, in fact, the earth has been and continues to be very dynamic and mobile throughout geologic time. Oceans transgress and regress over the land. Mountains rise, then erode and disappear. Continents move slowly around the face of the earth, sometimes colliding, forming larger landmasses and sometimes breaking apart, creating smaller landmasses.

In 1915, Alfred Wegener, a German scientist, published a book in which he proposed the <u>Theory of Continental Drift</u>. According to this theory, North and South America were once joined to Europe and Africa, forming Pangaea, and

---

Box 1. Continued

subsequently had "drifted" apart. His inspiration came from examining maps of the world. Wegener recognized the obvious fit between the continents if they were repositioned. Also, if these continents were once joined together, then geological features should match up from one side of the ocean to the other. Wegener spent a great amount of time and effort on this theory and succeeded in matching rock types, mountain ranges, fossils, and even living fauna and flora from one side of the Atlantic Ocean to the other. Thereby he <u>proved</u> these continents once were joined.

He called the ancient supercontinent "Pangaea", which means "all lands". Wegener's work was not eagerly accepted by the scientific community. It was not until the middle of the Twentieth Century that other scientists studying the sea floors and continents began to realize, as a consequence of their research, that Wegener's concept of moving continents was indeed correct. Today, we know that the surface of earth is composed entirely of a series of mobile fragments (or plates) which are continuously being created, moved and destroyed (consumed). The continents ride on top of these moving plates. This geologic process is called "*The Theory of Plate Tectonics*".

Geologists use the word "*tectonics*" to describe the processes that change the earth's *crust*, such as mountain building. It is derived from the Greek word, tekton, which is a carpenter or builder. In geologic terms, the word plate, refers to large fragments of the earth's crust.

Prior to the development of the <u>Theory of Continental Drift</u> and its successor, <u>Plate Tectonics</u>, most geologists believed the earth was static. Continents and oceans were believed to have been located at or near their present position more or less since near the creation of earth (approximately 4.6 billion years ago). The sedimentary record of rocks formed on continents since earth"s beginning is incomplete due to *weathering* and *erosion* processes which act on the surface of the land. In accordance with the belief the earth was static, geologists felt the oceans, being very old and permanent features of the earth, would contain a complete and uninterrupted record of sedimentary rocks from the oldest to the youngest. Geological discoveries which led to development of the <u>Theory of Plate Tectonics</u> literally turned the geological world upside down! Today, we know the ocean basins are the youngest major geographical and geological features on the surface of the earth. This is where new

Box 1. Continued

plates are created by rising basaltic magma from deep within the mantle. Based upon radiometric age dating of oceanic rock samples, the ocean basins do not contain any rocks older than 250 million years (quite juvenile considering the antiquity of the earth!). Instead, the continents contain the most complete sedimentary record (as incomplete as it is!) and the oldest rocks.

Depending upon how one defines the words "major" and "minor", the earth's surface is completely bounded by nine major plates and seven minor plates. These plates are continuously being created, moved, and destroyed. The uppermost layer of the plates is composed of heavy iron-rich *igneous rock*, called *basalt*. The continents, which are largely composed of lighter-density quartz-rich rocks, ride on the tops of basalt plates. The plates are bordered by three types of boundaries: *spreading centers*, *subduction zones* and *transform faults*.

From many geological and geophysical studies, scientists now theorize there are large *heat convection cells* that extend deep into the *mantle* of the earth. Heat convection cells serve to move heat from a hotter area to a colder area, and gravity is the mechanism that drives them. Hot material expands slightly and therefore its density per unit volume decreases. The same material, when cooled, contracts and its density, per unit volume, increases. The mantle of the earth contains very hot, iron-rich rocks called *ultramafics* which are under great pressure. Part of the mantle is not rigid and acts as a hot plastic that can slowly flow (*asthenosphere*). The hot, less dense plastic ultramafic rock rises to the surface of the earth, cools, then descends again, thereby transferring heat from the hot interior of the earth and dissipating it at the cool surface.

The mobile plates of the earth are believed to be the surficial manifestation of these heat convection cells. As hot, plastic ultramafic rocks slowly rise, partial melting occurs, creating basaltic *magma*, as the pressure on them lessens. The places where molten *basalt* reaches the earth's surface are called <u>spreading centers</u> or <u>spreading ridges</u>. These are long linear features. At spreading centers, hot molten *basalt* separates from the hot plastic ultramafics and rises to the surface, spreads laterally to either side, cools and solidifies. This process builds basaltic plates which move outward from the spreading center.

Box 1. Continued

How fast do the plates move? They move from 1 to 12 centimeters per year with an average of 1 to 2 centimeters per year. What does this mean to the people living on the surface of the earth? If one were to live all of his or her life in the same city and die at the age of seventy, during this time the city would have moved laterally approximately 140 centimeters (70 years times 2 centimeters/year) or approximately 4.6 feet! Not much to us in our lifetimes but try multiplying 2 centimeters/year times <u>millions</u> of years! In geologic time, the surface of the earth moves rapidly and changes dramatically. An idealized diagram of plate tectonics is shown in Figure 7.

How was the rate of movement of the plates determined? Basaltic plate rocks contain radioactive *minerals* which were created at the time of formation at the spreading center. Radiometric age dates may be obtained from samples taken from the plate rocks. Knowing the distance from the site of the sample back to the spreading center, one may divide this distance by the sample age and obtain the rate of movement (centimeters/year). Movements of the plates have also been measured from satellites.

We have discussed how new plate rocks are formed by hot molten *basalt* rising to the surface. However, we also know the earth is not increasing in size or mass. Therefore, if hot molten basalt is rising to the surface, then somewhere else cold, solid basalt must be going down. This indeed occurs in what are called <u>subduction zones</u>. Dense, basaltic plate rock is subducted back down into the earth's mantle. Spreading plates collide with each other and the plate that is the most dense will deflect downward (subduct) while the less dense plate will override it.

The third plate boundary is called a <u>transform fault</u>. If the rate of spreading of new basaltic rock is unequal from one side of the spreading ridge to the other, the ridge is stressed and is pushed toward the side of lesser spreading. Solid rocks are brittle and the stress is <u>transformed</u> along a fault into lateral displacement.

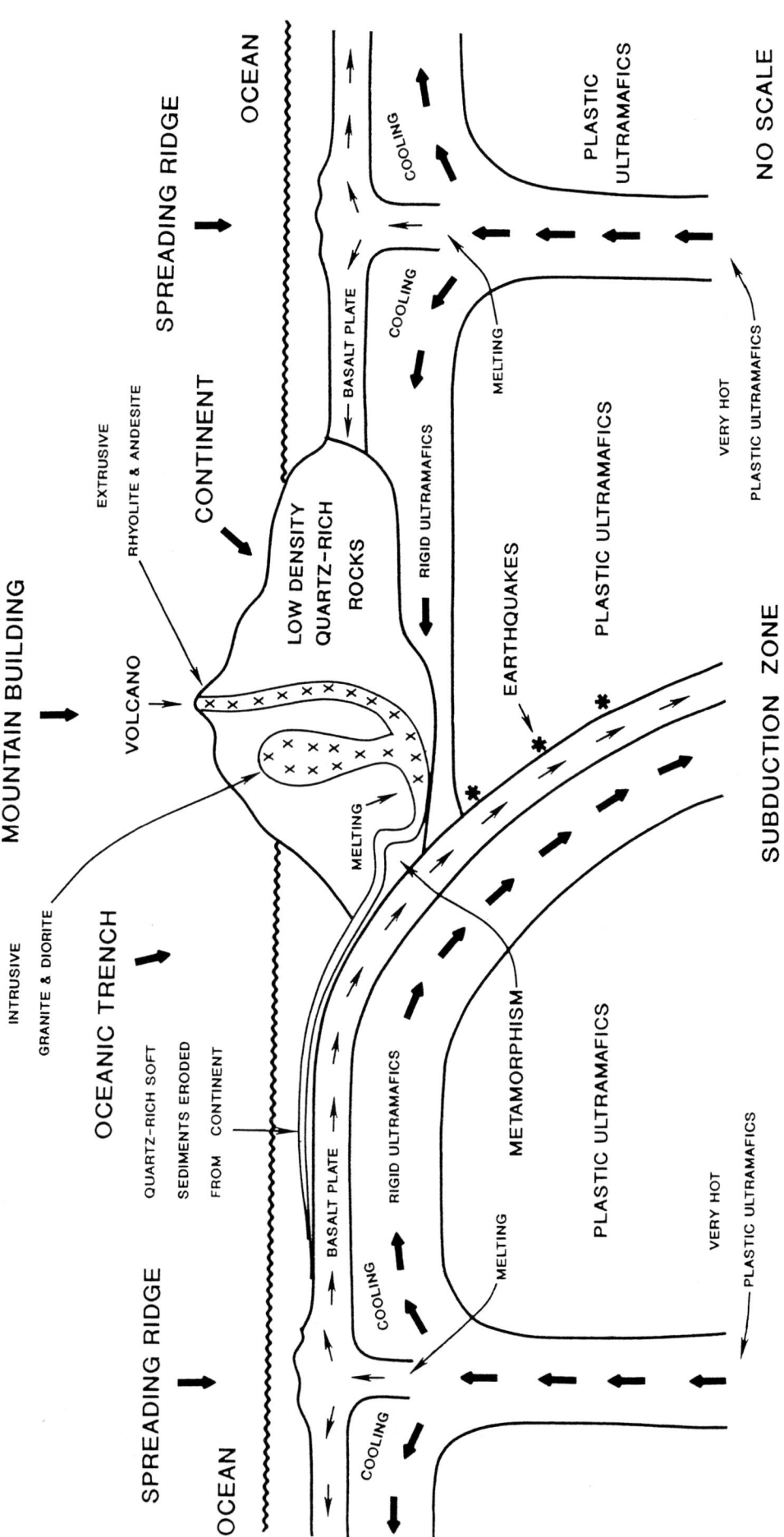

Figure 7. Plate Tectonics. Hot plastic ultramafics (very dense iron-rich rocks) rise slowly to the surface where they spread laterally. Near the surface as pressure is relieved, partial melting occurs and basalt separates out and extrudes, forming spreading ridges. The molten basalt then cools and solidifies, forming cold, solid basalt plates which migrate away from the spreading centers. Where plates collide, the one which is the most dense subducts under the other, forming subduction zones.

Sedimentary rocks are created by sedimentary processes of weathering, erosion, transportation and deposition. By these processes, rocks may be removed from some areas creating unconformities in the sedimentary record. Other areas may receive *sediments* by deposition resulting in new additional rocks. Common sedimentary rocks are *conglomerate*, *sandstone*, *shale*, *limestone* and *dolomite*.

A major factor in sedimentary processes is climate. Throughout geologic time, Colorado has experienced profound climatic changes. These have had a marked effect upon the landscape and the nature of the sedimentary rocks deposited. The climate of a given continental area is a function of many things, including circulation patterns of atmospheric and oceanic currents, elevation, location of geographic elements (such as mountain ranges) and latitudinal position. As a consequence of these factors, there are climatic belts which wrap around the earth, paralleling the equator. The tropical climatic zone lies on the equator. The semi-tropics border it on either side. Adjacent to these are two great desert belts, one in the northern hemisphere and one in the southern hemisphere. North and south of these lie the temperate climatic zones. The polar climatic zones cap the poles and extend out to the edge of the temperate zones.

The climate of a continental area then is dictated by its latitudinal position relative to the climatic belts and modified by local factors, as described previously. The nature of rocks created in these different climatic zones varies accordingly. For example, limestones form in shallow marine areas fringing continents in the tropical and semi-tropical zones. Glacial deposits occur in polar regions and high alpine areas. Sandstones are deposited in areas of the great desert belts. Climate clearly has much to do with the landscape and rocks formed in a given area.

Just to confuse the issue more, scientists now know that the continents have slowly migrated around on the surface of the earth, changing from one climatic zone to another. Ancient glacial features and deposits are found in areas which today have tropical to semi-tropical climates. Does this mean the earth suffered a massive, world-wide glaciation? Hardly! Those glaciers existed at a time when those land areas were located in a polar climatic zone, where glaciation is to be expected. The United States today is largely within the temperate climatic zone but it contains a majority of the world's coal deposits. Coals are created in coal-forming swamps of which there are few in the United States today. Coal-forming swamps are abundant in the tropics. Our continent gained its widespread Paleozoic age coal deposits at a time when it was geographically positioned on or near the equator.

The movements of continents through time have documented by intensive studies of the paleomagnetic properties of earth's rocks and by many other geologic studies. Based on these studies, the

North American Continent as we know it today has been located on or near the equator for a considerable portion of geologic time (at times the continent was even located in the southern hemisphere!). In those early times, Colorado was covered by shallow tropical oceans in which limestones and dolomites were deposited. During late Paleozoic and early Mesozoic time, the continent began to migrate into the northern hemisphere . As it passed through the great desert belt, Colorado was a desert characterized by extensive red mudflats (evidenced by red mudstones) and areas of great sand dunes (evidenced by thick *eolian* sandstones). It is easier to understand the climatic changes caused by plate migrations if the "dance of the continents" is taken into account. For additional information on climatic changes and the movements of continents, see Box 2.

PRECAMBRIAN TIME:

The story of *geology* in Colorado begins with the oldest rocks, those that were formed in the Precambrian. The Precambrian spans four billion years of the history of the earth.

Precambrian rocks of Colorado have localized formation names or even in some places no names at all. In these unnamed areas, rocks are simply referred to as "Precambrian". The Precambrian is composed primarily of *metamorphic gneiss* and *igneous granite*, with lesser amounts of *schist, phyllite, slate, quartzite* and rarely *marble*. Some areas contain locally abundant white to pink *granitic pegmatites* which have intruded into metamorphic rocks such as gneiss. In the extreme northwest part of the state, late Precambrian age *sedimentary sandstones* are exposed in the Uinta Mountains.

The ages of the metamorphic rocks are based upon radioactive age dating and most of the gneisses are about 1.7 to 1.8 billion years old. Granites were intruded from below into overlying rocks (usually metamorphic rocks) in three separate episodes of 1, 1.4 and 1.7 billion years ago. The oldest rocks of the world are 4.3 plus billion years old. Therefore Colorado's Precambrian rocks are relatively "young." The Precambrian rocks of Colorado are exposed in the higher parts of the mountains where they have been exposed by erosion.

BOX 2. "DANCE OF THE CONTINENTS"

The continents have moved back and forth, sometimes with a circular motion, in response to the ceaseless movements of the earth's crustal plates as they are created and destroyed by spreading ridges and subduction zones (see Box 1). As a consequence of this movement, continents sometimes move through different climatic zones. At least in part, Colorado's climate has been affected by these changes in latitude throughout geologic time.

In the late Precambrian and during most of the Cambrian, Ordovician, Silurian, and Devonian periods, North America was positioned south of the equator. Thus Colorado was in the tropics with a tropical climate (See Figure 8).

In early Pennsylvanian time, the continent began to move back to the north, crossing the equator. By the middle Pennsylvanian, it was in the Northern Hemisphere (still within the tropics). The giant supercontinent, *Pangaea*, began to coalesce during the Pennsylvanian and persisted into the Early Jurassic, when it began to break up.

As time passed, the North American Continent continued to move north. In the late Pennsylvanian, Permian, Triassic and early Jurassic, much of it was in the desert zone north of the equator. As today, there are two great desert belts around the world centered between 20^0 and 35^0 Latitude north and south of the equator. In these zones, warm dry air descends, creating extensive deserts. As Colorado drifted through the Great desert belt, it experienced arid conditions. At times, it was covered in part by great sand dune fields, similar to the Sahara Desert of today. At other times, extensive dry red mudflats blanketed the terrain.

In late Jurassic time, North America moved to the northwest, leaving the desert zone. It entered a more humid climatic zone which was probably semi-tropical.

By Cretaceous time, the continent was very near its present latitude and it has been moving westerly ever since. The climate of Colorado remained warm (borderline semi-tropical to temperate) through the rest of the Cretaceous and the Tertiary. With the advent of the Pleistocene *glaciation*, the area cooled drastically.

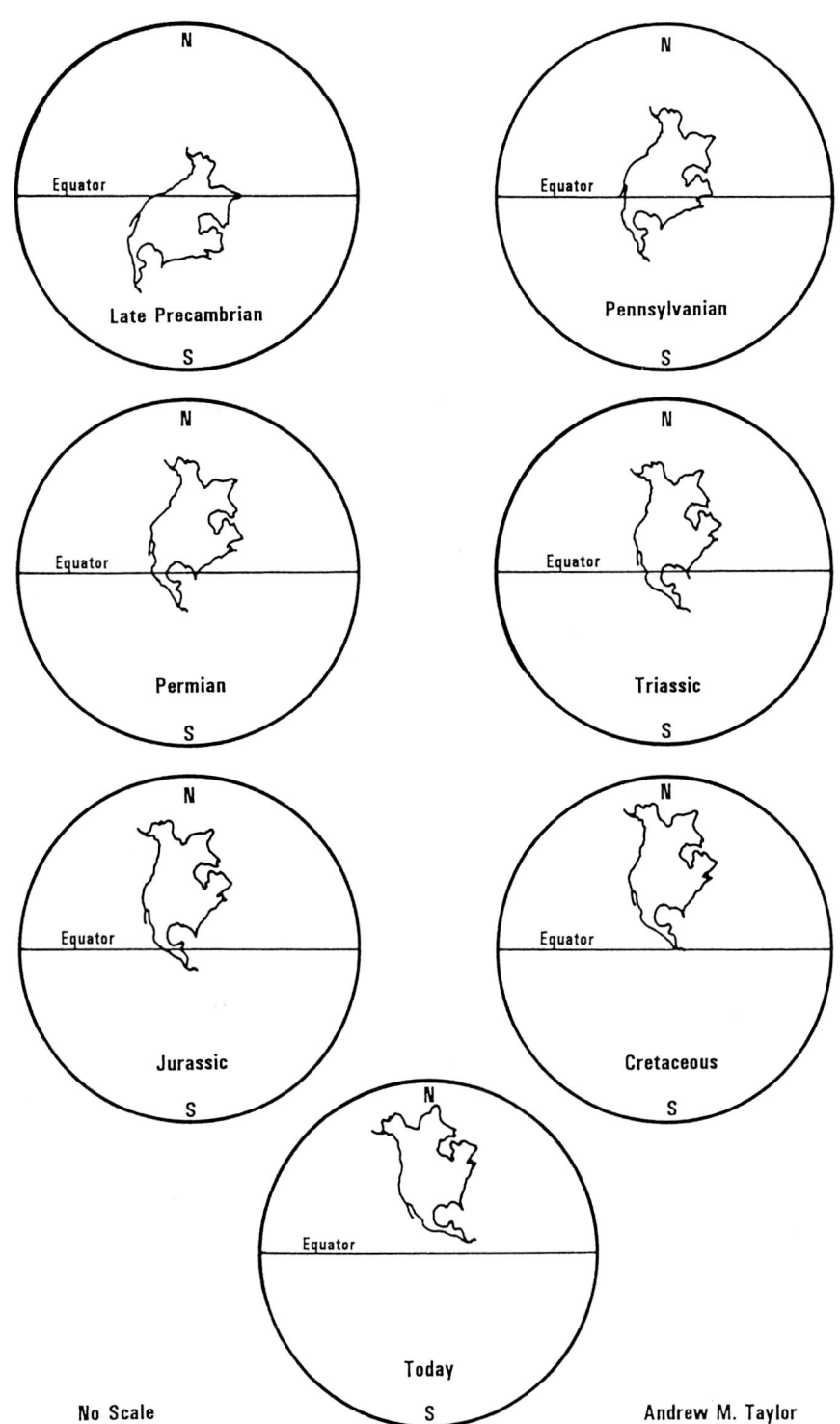

No Scale

Andrew M. Taylor

Figure 8. Migration of the North American Continent:
Studies of *paleomagnetism* allow the reconstruction of continental migration with respect to latitudes. In this figure, the shape of the North American Continent is shown as it appears today. In earlier geologic periods, it actually had a different configuration and was smaller.

(Discussion of Figure 8 continued on next page)

21

Late Precambrian: The North American Continent was positioned south of the equator and was rotated approximately 90^0 from its present-day orientation. The continent remained largely in the southern hemisphere through the Devonian.

Pennsylvanian: During the Mississippian Period, the continent began to move to the north, crossing the equator. As shown in the diagram, by Pennsylvanian time, it was about two/thirds into the northern hemisphere. Also during the Pennsylvanian, North America began to collide with Africa and Europe, an early step in the creation of the supercontinent, *Pangaea* (not shown). The Appalachians and the Ancestral Rockies were formed at this time.

Permian: The continent continued to move to the north and extended from the equator to nearly 60^0 north latitude. Part of the continent lay within the great northern desert belt between 20^0 and about 35^0 north latitude. The formation of Pangaea was completed.

Triassic: The slow march of the continent continued north and progressively passed through the great desert belt.

Jurassic: During early Jurassic, North America was still moving through the desert belt. By late Jurassic, it had moved north of the arid region and into a more humid region. Today we call this the temperate zone. The breakup of Pangaea began in the Jurassic.

Cretaceous: By the end of the Cretaceous, the North American Continent had moved approximately as far north as it is today. Near the end of the Cretaceous, the direction of plate movement changed to westerly as the North American plate began colliding with the Pacific oceanic plate. The building of the modern-day Rocky Mountains began.

ROCK TYPES OF THE PRECAMBRIAN:

METAMORPHIC:

Metamorphic rocks are rocks which have been changed by heat and pressure into new rock types <u>without</u> being heated to the melting point. Therefore, metamorphic rocks have "*parent rocks*", the original rock before *metamorphism*. It is very meaningful to geologists to determine the parent rocks because this reveals still earlier geologic history.

Origin	Rock	Parent Rock
Metamorphic	gneiss	marine graywackes, shales and felsic to mafic marine lavas
	schist	largely marine sedimentary rocks and some volcanic tuffs
	quartzite	marine quartz sandstone
	marble	marine limestone

Table 1. Common Precambrian Metamorphic Rock Types of Colorado and their Parent Rocks (Tweto, 1980, p.37, 41).

IGNEOUS:

Most of the *intrusive igneous rocks* of Colorado are *granites*, although some *diorites* and rarely *gabbros* occur.

Origin	Rock	Mode of Emplacement
Igneous Intrusives	granite	intruded upward into older overlying metamorphics
	pegmatite	intruded into metamorphics

Table 2. Common Precambrian Igneous Rock Types of Colorado and their Mode of Emplacement.

SEDIMENTARY:

Unmetamorphosed *sedimentary rocks* of Precambrian age are uncommon in Colorado but late Precambrian pebbly *sandstones* do exist in the far northwestern corner of the state.

EASTERN COLORADO:

Idaho Springs Formation: Principally *gneiss* (See Figure 9) with lesser amounts of *schist* and uncommonly *marble*. The Idaho Springs is 1.7 billion years old and this name is used widely in the Front Range of Colorado. In some areas, these rocks are simply called *metamorphics*.

Pikes Peak Granite: Composed of *granite* 1 billion years old. The *batholith* (a very large body of granite) is located in the Pikes Peak area (See Figure 10).

Figure 9. Gneiss. Boulder of gneiss in Medano Ck., Great Sand Dunes National Monument. Note the typical layering of light and dark layers. Boulder is approximately 18 inches long. Photo by Lennell Gallegos.

24

Figure 10. Pikes Peak Granite. The Pikes Peak Granite is exposed in a large area located in the mountains west of Colorado Springs. Granite, being of a relatively homogeneous structure and composition, typically weathers to a highly recognizable landform of rounded surfaces and boulders. Picture taken on U.S. Highway 24 west of Colorado Springs between Divide and Florissant.

25

WESTERN COLORADO:

Metamorphic Gneisses and Schists (unnamed) are abundant basement rocks. Their character and age (1.7 to 1.80 B.Y.) are similar to those of the Idaho Springs Formation.

Uinta Mountain Group: Composed of light to dark red pebbly sedimentary *sandstone* with some thin beds of *shale*. These sandstones are approximately 24,000 feet thick and are believed to be 950 million years old at the top and 1.4 billion years old at the base. The sandstones are stream deposits and, together with the dominant red color, clearly are derived from *terrigeneous* (land) sources. During late Precambrian time, a large continent lay to the north of Colorado and it appears that the southernmost flank of this continent extended down into the northwestern part of the state while the rest of Colorado was *marine*. This continent to the north presumably was the source of the sandstones of the Uinta Mountain Group.

Uncompahgre Formation: The rocks of the Uncompahgre have undergone a modest amount of *regional metamorphism* and are now composed of *slate*, *phyllite* and *quartzite*. The *parent rocks* were *shale* and *sandstone*. The rocks of the Uncompahgre Formation are older than those of the Uinta Mountain Formation, somewhere between 1.4 and 1.7 billion years old.

INTERPRETATIVE ORIGIN OF PRECAMBRIAN ROCKS:

Geologists know less about the Precambrian than the Paleozoic, Mesozoic and Cenozoic Eras. Partly this is because Precambrian rocks are the *basement rocks* and have less exposure at the earth's surface compared to younger rocks. Furthermore, as rocks become older, there is more time for geologic processes to alter and change them, obscuring their original character. As a general rule, the older rocks are, the less geologists can learn about them. What we know about the Precambrian is what the rocks tell us.

Precambrian rocks of Colorado:

1. Were largely formed by widespread, intense *metamorphism* and *igneous intrusive* activity. *Sedimentary rocks* formed by sedimentary processes are not as abundant.

2. Are largely *gneiss* and *granites*, with some *sedimentary sandstones* in the northwestern part of the state.

3.   That are *metamorphic* have *parent rocks* which were deposited on the bottoms of oceans (*marine arkosic sandstone* and marine *extrusive basalt*).

4.   Contain many areas of intensely folded *gneiss*, suggesting extreme pressure and temperature.

Based upon the previous observations, the following interpretation is presented:

1.   During Precambrian Time, the earth was hotter, as suggested by the widespread metamorphism and igneous activity.

2.   For the same reasons, *heat convection cells* within the earth (as described in Box 1) moved much faster and were probably more numerous.

3.   Continents during early Precambrian Time probably were much smaller, possibly only *island arcs* (such as the Aleutian Islands of Alaska). In late Precambrian Time, continents grew by *accretion* (where sediments are scraped off a subducting plate and added to a continent. See Figure 7 of Box 1) and began to assume their present character.

4.   Low density, quartz-rich rocks were recycled very fast by subduction, accretion and igneous intrusion. Most of the world's quartz is in gneisses and granites.

5.   Summary of probable geologic processes involved:

     a.   Weathering and erosion
     b.   Marine sedimentation
     c.   Subduction and accretion
     d.   Metamorphism
     e.   Igneous activity (intrusion of granitic magma into the upper crust of the earth)
     f.   Uplift (mountain building)

Large amounts of metamorphic rocks were created in this manner. Some were carried deeply enough down *subduction zones* to melt and become *igneous magma*. Later the magma intruded the overlying metamorphic rocks and formed granites and *granitic pegmatites* (tabular bodies of granite with large mineral grains).

Therefore, during early Precambrian time, a considerable portion of the quartz-rich rocks of the earth ended up as metamorphic rocks (principally gneisses) and igneous granites. In the late Precambrian, the continents began to form. Some areas of the earth have abundant Precambrian-age sedimentary rocks (largely sandstones and shales) which "survived" and were not recycled by

27

the processes of *Plate Tectonics* into metamorphic and igneous rocks.

"Basement rocks" of continents (and Colorado) are therefore largely Precambrian gneisses and granites. Beneath the "basement rocks" of continents lie *ultramafics* (rocks composed of magnesium and iron rich minerals such as olivine and augite) of the rigid uppermost layer of the *mantle*.

The total thickness of Precambrian metamorphic rocks in Colorado is not known. These rocks (plus granites) may extend down to the mantle at the bottom of the continent.

To recap, the creation of Precambrian rocks appears to be a story of plate tectonics acting at a very fast pace on a hotter earth. Colorado during this time was largely under an ocean where subduction zones were active. To the northwest, a continent apparently existed during late Precambrian Time and this portion of Colorado probably was above sea level.

Life on earth originated in the Precambrian approximately 3.5 *BYA*. The early organisms were simple consisting of a very few cells. As time passed, these evolved into more complex organisms. They were soft-bodied and lacked hard parts (shells or skeletons). Consequently, Precambrian fossils are sparse and geologists know very little about these early forms of life.

PALEOZOIC ERA

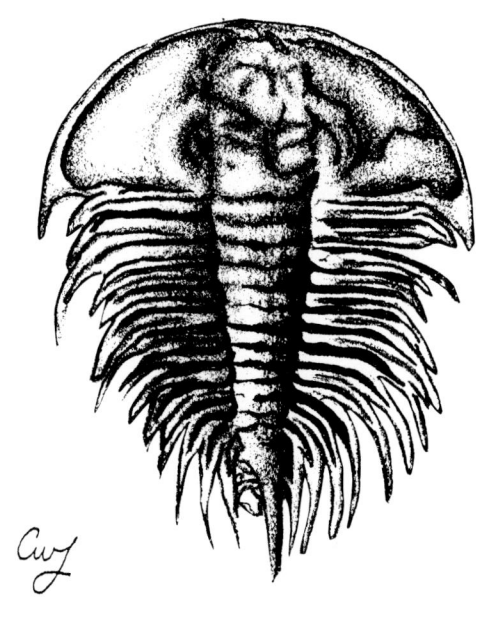

Cambrian Trilobite

Shellfish evolved in the very late Precambrian and their fossils suddenly appear in the *sedimentary rocks*. The appearance of these fossils is used to mark the end of Precambrian Time and the beginning of the Paleozoic Era. Shellfish extract calcium (Ca^{++}) and carbon dioxide (CO_2) from seawater and build *calcite* ($CaCO_3$) shells. Some plants also evolved which built plant parts out of $CaCO_3$. Upon death of these organisms, the calcite shells and plant parts provided *calcareous* sediment (soft mud) which was later lithified (changed by depth of burial, compaction and cementation) into hard *limestone* and *dolomite*. Geologists study modern depositional environments in order to learn more about older rocks which were created in similar situations. For example, studies of extensive areas of shallow tropical marine waters in the Bahamas have been made. These areas have high populations of shellfish and other marine organisms that build shells made of calcite. The sea bottom in these shallow areas is composed of soft carbonate mud. The mud contains abundant fragments of shells which have broken up. Limestones are composed of hard cemented calcareous mud containing broken fragments of shells. Studies such as these reveal to geologists how rocks are formed. From the beginning of the Paleozoic Era through the Mesozoic and Cenozoic eras to the present, these organisms that build shells and skeletal parts composed of $CaCO_3$ have tied up enormous quantities of CO_2 in the resulting limestone and dolomite. For a fascinating story of limestones, see Box 3.

Throughout much of Paleozoic time, The North American Continent (and Colorado) was located near the equator in a tropical environment. The earth was warmer then and there were no polar ice caps. Consequently, the sea level was higher and about two/thirds of the continent was covered by shallow tropical seas. Extensive Paleozoic limestone and dolomite rocks were deposited on these continental shelf areas.

BOX 3. LIMESTONES

Limestones are ugly rocks. They are grey,...and ugly! Limestones may only be regarded otherwise when they contain beautiful fossils, which they sometimes do. But, limestones are special rocks, in fact, they are wonderful rocks. They are literally the rocks of life. Indeed, if limestones did not exist in abundance on earth, neither would we! Limestones are intimately involved in the evolution of the earth's atmospheres. A rather important consideration in as much as we have to have one to breathe. This the story of limestones.

DEVELOPMENT OF THE EARTH'S ATMOSPHERES

ASSUMPTIONS:

In all likelihood, earth originally had an atmosphere compatible with the nature of the universe (hydrogen and helium constitute 99% of the total mass of the universe).

Today, earth's atmosphere, however, contains only minute amounts of pure hydrogen and helium, therefore earth must have lost its first atmosphere.

Earth's gravity holds its atmosphere. Escape velocity (the speed to escape Earth's gravitational field) exceeds 7 miles/second. Only the lightest atoms (hydrogen and helium) occasionally reach such velocities.

1.   FIRST ATMOSPHERE: (more than 4 *B.Y.A.*)

Helium is very abundant in the universe but extremely rare in our atmosphere. Hence earth must have lost its first atmosphere of hydrogen and helium to space. To lose it, the earth must have been hot enough to give gases the necessary escape velocity.

2.   SECOND ATMOSPHERE: (about 4 *B.Y.A.*)

It is believed that *volcanoes* formed the second atmosphere. *Volcanic* gases emitted include H_2O (water), CH_4 (*methane*) and NH_3 (*ammonia*). As earth continued to cool, water vapor in the atmosphere condensed and filled the ocean basins and therefore earth now had oceans. At this time, the atmosphere was composed mainly of ammonia and methane with some remaining water vapor.

Box 3. Continued

3.   **THIRD ATMOSPHERE:**   (between 4 *B.Y.* and 3.5 *B.Y.A.*)

The second atmosphere was exposed to millions of years of solar radiation. *Ultraviolet* light decomposes ammonia (NH_3) into hydrogen and nitrogen and methane (CH_4) into hydrogen and carbon. The remaining water vapor (H_2O) in the atmosphere also decomposed into hydrogen and oxygen. The hydrogen thus liberated, escaped into space, leaving nitrogen, oxygen and carbon. The carbon and oxygen combined to form carbon dioxide (CO_2). Hence the third atmosphere was composed of nitrogen and carbon dioxide.

4.   **FOURTH ATMOSPHERE:**   (beginning 3.5 *B.Y.A.*)

Life had originated late in the third atmosphere, with plants being the first to evolve in the oceans. Using *photosynthesis*, they extracted carbon dioxide from water and released oxygen into the ocean. Carbon dioxide easily dissolves into water from the atmosphere. Therefore the atmosphere and oceans act as a common reservoir for carbon dioxide. In these early times, the oceans held vast quantities of dissolved iron. As the newly evolved plants began to release oxygen into the oceans, the iron was oxidized, forming the iron *minerals*, *hematite* and *magnetite*. The heavy minerals fell to the sea floors and created the banded iron ores. Today, these are our primary ores of iron. When the ocean became saturated with oxygen, oxygen began to escape slowly into the atmosphere. For the very first time, free oxygen accumulated in the atmosphere. Later, plants moved onto the land. These land plants took carbon dioxide directly from the air and returned free oxygen to the atmosphere. Some plant materials were later transformed into coal, oil and gas. Carbon in these materials has been stored for long periods of geologic time.

As time passed, *shellfish* and other organisms evolved in the oceans. These organisms extracted carbon dioxide and calcium from seawater. They built shells and structures composed of calcium carbonate, ($CaCO_3$, which is initially the mineral aragonite which later changes to *calcite*), as per the following equation:

$$Ca^{++} + CO_2 + H_2O = CaCO_3 + H_2$$

As can be seen in this equation, every molecule of *calcite* contains one molecule of carbon dioxide.

Box 3. Continued

These shells and skeletons of calcite ultimately accumulated on the seafloor to form *calcareous* mud. Frequently, the mud was compacted and transformed into *limestones* and *dolomites*. Enormous quantities of carbon dioxide are stored in the world's abundant *sedimentary* limestone and dolomite rocks

These profound biologic and geologic processes virtually cleansed carbon dioxide out of the third atmosphere (and oceans) and stored it in biomass, fossil fuels and *calcareous* rocks (limestone and dolomite). This process yielded our fourth atmosphere composed of principally nitrogen and oxygen (Haber, 1969, p. 46).

Scientists have estimated 20,000,000 gigatons (1 billion metric tons) of carbon dioxide are stored in carbonate rocks. All other carbon dioxide in the atmosphere, oceans, biomass, soil and fossil fuels is but a trace compared to this (see U.S. Department of Energy, 1980).

In summary, the origin of life itself has greatly modified the earth. It has been instrumental in cleaning all but a trace of carbon dioxide out of the atmosphere, accompanied by the release of free oxygen. Enormous quantities of carbonate rocks were created from the shells and skeletons of organisms. These rocks now contain virtually all of the earth's carbon dioxide. If they were to decompose, the carbon dioxide would be released and our atmosphere would return to a composition of principally carbon dioxide. The next time you encounter a limestone, give it a friendly pat. It is truly a wonderful rock, even if it is ugly!

CAMBRIAN PERIOD: 570-500 *M.Y.A.*

EARLY CAMBRIAN:

No rocks of Early Cambrian age are present in Colorado. Instead, there is a large *unconformity* on top of Precambrian rocks. Such large unconformities are usually interpreted as meaning that the land during this time was *emergent* above sea level. There was no deposition, instead, there was *erosion*. However, it is possible that these rocks were deposited but later uplifted and erosion removed them at the end of Early Cambrian time.

LATE CAMBRIAN:

Central Colorado was covered by a tropical ocean depositing *beach sandstones* as it spread from west to east.

Sawatch Formation: *Quartz sandstone*, often altered to a *quartzite*. *Brachiopod* fossils are found in the rocks.

Peerless Formation: *Marine* glauconitic *sandstone*. The *mineral glauconite* is green in color and is derived chemically from ancient fish fecal pellets that were incorporated in the sediments. Immediately west of Colorado Springs, the Peerless contains zones of a strong green color from the glauconite.

ORDOVICIAN PERIOD: 500-435 *M.Y.A.*

The tropical sea deepened and widened over Colorado and carbonate sediments were deposited which later lithified into hard rocks.

Manitou Formation: *Cherty limestones* and *dolomites* containing sparse *shellfish*, snail, *echinoderm* and *sponge* fossils.

Harding Sandstone: *Sandstones* containing fossil armor plates from primitive, jawless fish (Agnathids). After the Manitou was deposited, the sea became shallower and a source of terriginous sand developed.

Ordovician Brachiopod

Fremont Dolomite: Massive gray *crystalline dolomite* (in places a *limestone*) containing *mollusk*, *brachiopod* and *coral* fossils. Coral fossils are common. After the deposition of the Harding Sandstone, the sea deepened again and deposition of carbonate sediments resumed.

33

SILURIAN PERIOD: 435-410 *M.Y.A.*

Silurian rocks are very rare in Colorado.   In fact, until 1978, it was believed that Colorado and several surrounding states were *emergent* and no rocks were deposited during the Silurian. However, in 1978, Silurian rocks were found as broken blocks of *limestone* in *igneous pipes* (special types of volcanoes called *diatremes*) located in the Front Range in far north central Colorado and adjacent portions of Wyoming.   The igneous pipes were intruded during the Middle Devonian Period.   The presence of these Silurian limestone blocks in the igneous pipes proves Silurian rocks were present in this area.   Shallow Silurian seas did cover Colorado. Silurian limestones were deposited but were later removed by erosion over the state.   Some of the diatremes have been found to contain diamonds and are therefore called "diamond pipes".   For a more detailed discussion of the diamond pipes, see Box 4, entitled "Diamonds in Colorado".

BOX 4.   DIAMONDS IN COLORADO

Diamonds in Colorado?   The first association of diamonds with the state of Colorado was somewhat embarrassing.   This event has been called "The Great Diamond Hoax".   In the spring of 1872, two prospectors named Philip Arnold and John Slack successfully perpetrated an elaborate gemstone hoax on a number of wealthy San Francisco businessmen.   The two miners appeared at the Bank of California in San Francisco with a bag of uncut diamonds, rubies, sapphires, emeralds and garnets which they claimed they had found at a site in Arizona.   This site was later found to be in northwestern Colorado.   The president of the bank, William C. Ralston, was very interested in their claims and he struck a deal with Arnold and Slack.   He formed a company and raised two million dollars from 25 wealthy investors (one of these investors was General George McClellan, the Civil War General).

To his credit, Ralston performed considerable due diligence in an effort to determine the veracity of the claims of Arnold and Slack.   Unfortunately at this time, there were very few people in the United States who knew anything about the geology of gemstones.   Ralston had no connections with those who did, until it was too late.

Ralston initially sent two of his associates with the miners to the site.   This involved a 36 hour train ride and four days blindfolded on mules to the purported mine.

Box 4. Continued

There, they easily found many gemstones and returned triumphant to San Francisco with a bag of them. Jewelers in San Francisco appraised a parcel at $125,000.

Ralston sent the same parcel to New York City, where Charles Tiffany, the most famous jeweler in the United States, appraised them at $150,000.

Ralston then hired a well experienced and highly respected mining engineer to go (blindfolded as usual) and evaluate the gemstone deposit. The engineer declared the deposit to be real and valuable (this from a man who had never seen a gemstone deposit before in his life!).

Convinced, Ralston and his investors bought out Arnold and Slack for $600,000! In 1872, this was a very large sum of money. At that time, gold was valued at $20 per ounce. Today, gold has a value around $300 per ounce. In terms of $300 gold, this is equivalent to $9,000,000 in current dollars! The $600,000 was paid to Arnold (Slack had mysteriously disappeared earlier) and Arnold left town with the money. It is unknown whether Slack ever received any of the money.

At this time, a geologist by name of Clarence King became involved. Clarence had earlier led the Fortieth Parallel Survey for the government. The Survey prepared maps and detailed reports on everything from minerals to vegetation and weather along a 100 mile wide by 1,000 mile long portion of the Fortieth Parallel. The discovery of the fabulous gemstone deposit was widely reported in the nation's press. King, upon learning of it, was deeply troubled for more than one reason. He realized the site had to be within the area of his survey and it would be extremely damaging to his career if he had missed such a discovery. In addition, King knew that the geological conditions for the formation of diamonds, rubies and sapphires, emeralds and garnets are vastly different. All of them never occur together, so he strongly suspected the existence of a fraud.

King visited the mining engineer who had evaluated the deposit. From his description of the area, King believed he knew where the site had to be. He then made his own trip and succeeded in finding the gemstone deposit. King determined it had been salted and was indeed a hoax. He then journeyed to San Francisco and delivered the unpleasant news to Ralston.

Box 4. Continued

Ralston was, of course, exceedingly distressed at this news. He repaid the investors out of his own pocket and hired detectives to track down the swindlers. The detectives found Arnold in Kentucky where he had used the money to buy his own bank. Kentucky refused to extradite Arnold. However, Arnold did return $150,000 to Ralston in exchange for immunity from further prosecution. It was later learned that Arnold and Slack had made a trip to London and Amsterdam where they purchased the poor quality gemstones used in the scam. It was never learned where the money to buy the gems came from. A year later, Arnold was killed by a rival bank owner. Slack was never found.

Charles Tiffany was very humiliated by the unexpected turn of affairs. He admitted that he was inexperienced with uncut gemstones and had overestimated their value. In 1875, Ralston's Bank of California went under and Ralston's body was found floating in San Francisco Bay.

In 1872, the borders of western states were ill-defined. Today, it is known the locality of the salted gem mine is within Moffat County in northwestern Colorado near the borders of Wyoming and Utah. Interestingly, a nearby mountain is named Diamond Peak and a drainage on the northeast side of the mountain is called Diamond Wash Gulch (See Figures 11 and 12).

As a consequence of his work in revealing the scam, Clarence King, the geologist, became well known. He later became the first Director of the U.S. Geological Survey and one of the most famous and prestigious geologists in America. Most of the other persons involved in the "Great Diamond Hoax" either were damaged by it or came to a tragic end. King was the only one benefiting from it.

After the "Great Diamond Hoax" and until recently, the only diamonds in Colorado were found in jewelry stores. In 1976, scientists at the U.S. Geological Survey were preparing a sample from an unusual rock found at a site in southern Wyoming. Something in the rock scored a grinding plate. This was, to say the least, surprising. The only natural mineral hard enough to cause such scratches is diamond! When examined under a microscope, a small diamond was found. Within a short period of time, diamonds were discovered in similar rocks at sites south of the Wyoming border in Colorado. Thus real diamonds do occur naturally in Colorado!

Figure 11. Diamond Peak. As viewed to the west from County "10N Road", the peak on the far horizon bears the name "Diamond Peak" on USGS topographic maps. It is composed of Tertiary conglomerates. Presumably, it obtained its name from the "Great Diamond Hoax". Diamond Peak is located on the northeast side of Cold Spring Mountain, part of the Uinta Mountains.

37

Figure 12. Diamond Field Draw and Diamond Field. This small drainage from the northeast side of Diamond Peak bears the name of Diamond Field Draw. The flat treeless area on the upper left side of the photo is called Diamond Field. The site of the Great Diamond Hoax is either in the field of view or nearby. The area, which is located in Moffat County, may be reached from County "10N Road".

38

Box 4. Continued

This revelation was the result of a series of geological discoveries. Geology is very much an observational science. Many times, important geological discoveries are made by accident. Geologists are prone to spend a lot of time out studying rocks. This frequently yields the unexpected. Such is the story of diamonds in Colorado. It involved a series of events which started with the discovery of Silurian-age *limestone* blocks. This led to the identification of *kimberlites* (more about these later). Finally the kimberlites were found to contain diamonds.

SILURIAN LIMESTONES:

Prior to 1976, the common belief among geologists was that there were no Silurian age rocks present in Colorado. Simply stated, such rocks had never been found in Colorado and surrounding states. This large area was believed to have been above sea level during Silurian time and no *marine* sediment had been deposited.

In 1960, two geologists named C.S. Ferris, Jr. and C.A. Aultman unexpectedly found Silurian limestones containing well preserved *corals* and *brachiopods* (similar to *clams*) in the Sherman Mountains of Wyoming's Laramie Range. These limestones were situated in an area composed of Precambrian-age rocks, a most unlikely place for Silurian marine limestones to be present. In 1964, a geologist by the name of M.E. McCallum found similar Silurian limestones in an area of Precambrian rocks on the Sloan Ranch in Colorado (See Figure 13).

Geologists then puzzled over the meaning of these limestones. Did this mean Colorado and the surrounding area <u>were</u> covered by Silurian seas and the resulting limestones later were largely removed by *erosion* (an *unconformity*)? The answer to this enigma did not become apparent until later.

Later in the same year (1964), McCallum recognized *serpentine*-rich *breccia* rock samples from the Sloan Ranch location as being *kimberlite*. He therefore identified the first *diatreme* which became known as the Sloan Diatreme (See Figure 14). The Ferris and Aultman location was then also recognized as a diatreme (diatremes are a special type of *volcano* which will be described in detail at the end of this section).

Box 4. Continued

Subsequently, other diatremes were discovered, but none were believed to contain diamonds. In 1976, kimberlite samples from the Schaffer #3 Diatreme located in Wyoming were submitted to the U.S Geological Survey. Diamonds were found in them as recounted previously. Within a short time, diamonds were recognized in the Sloan Diatreme in Colorado. Other diatremes in what is now referred to as the "State-Line District" also have been found to contain diamonds.

After discovery of diamonds in the kimberlites, a small "diamond rush" occurred. By 1997, approximately 100 occurrences (personal communication from Dan Hausel, State Geologist of Wyoming) of kimberlite have been located. Not all of these contain diamonds.

SO WHAT ARE KIMBERLITES AND DIATREMES?

Basically, *diatremes* are small *volcanic* pipes emplaced in rocks by gaseous explosions. They are filled with angular broken fragments called *breccia*. Diatremes that contain kimberlite have very deep roots (greater than 150 kilometers) into the earth's *mantle*. Kimberlite is formed from *magma* and rock fragments brought up in the diatreme from deep sources in the mantle. Generally, geologists believe kimberlite magmas result from partial melting of peridotite material 150 to 350 kilometers below the earth's surface. In the kimberlite magma, there are large amounts of dissolved gases (CO_2 and H_2O) under great pressure, plus *xenoliths* of peridotite and other rock types. Sometimes, *xenocrysts* of *minerals* such as diamonds are included. The resulting solid rock, *kimberlite*, is commonly composed principally of *olivine* or *serpentine*, with lesser amounts of *phlogopite*, *chrome diopside*, *calcite*, red pyrope *garnet*, *ilmenite*, *spinel* and other minerals. Diamond is a rare constituent and is not found in all kimberlites. Kimberlites also contain fragments (xenoliths) of various rocks penetrated by the volcano as it blew to the surface (See Figure 15).

The gas involved in the creation of diatremes is believed to be principally carbon dioxide (CO_2). Groundwater (which could vaporize) may also be a factor as the eruption nears the surface. Kimberlitic diatremes have a characteristic carrot shape with a root zone which extends into the depths of the mantle. The gaseous eruption blows out the surface, creating a small crater ringed by ejecta deposits. Diamonds are formed in the mantle and sometimes are brought up by the eruption. Known diatremes in the

Box 4. Continued

world range from very old (1.5 billion years), to as young as possibly a few million years. None are known to have occurred during historical times. Of those containing diamonds, relatively few have enough diamonds to be economic.

Figure 13. Limestone Boulder in the Sloan Diatreme. In the foreground, a large boulder of limestone is exposed. Some of the boulders in the Sloan Diatreme (located in the State Line District) are Ordovician in age. This is based upon their fossil content. Other Sloan boulders may be of Silurian age as indicated by their fossils. These limestone boulders fell or moved down into lower depths of the diatreme at the time it was formed. Contrary to prior belief that Silurian rocks were not deposited in Colorado, Silurian rocks in diatremes of the State Line District provide evidence that marine Silurian limestones were deposited over this area.

Figure 14. The Sloan Diatreme #1. Grayish kimberlite outcrops adjacent to Precambrian granites on the left. This is part of the Sloan Diatreme #1, located in the State Line District. It is on private property and is not open to the public. Two attempts to mine it for diamonds were unsuccessful. The surface of the diatreme has been revegetated and underlies the foreground. Scattered over the outcrop are participants of a Rocky Mountain Association of Geologists field trip conducted by W. Dan Hausel, Senior Economic Geologist, Wyoming State Geologic Survey.

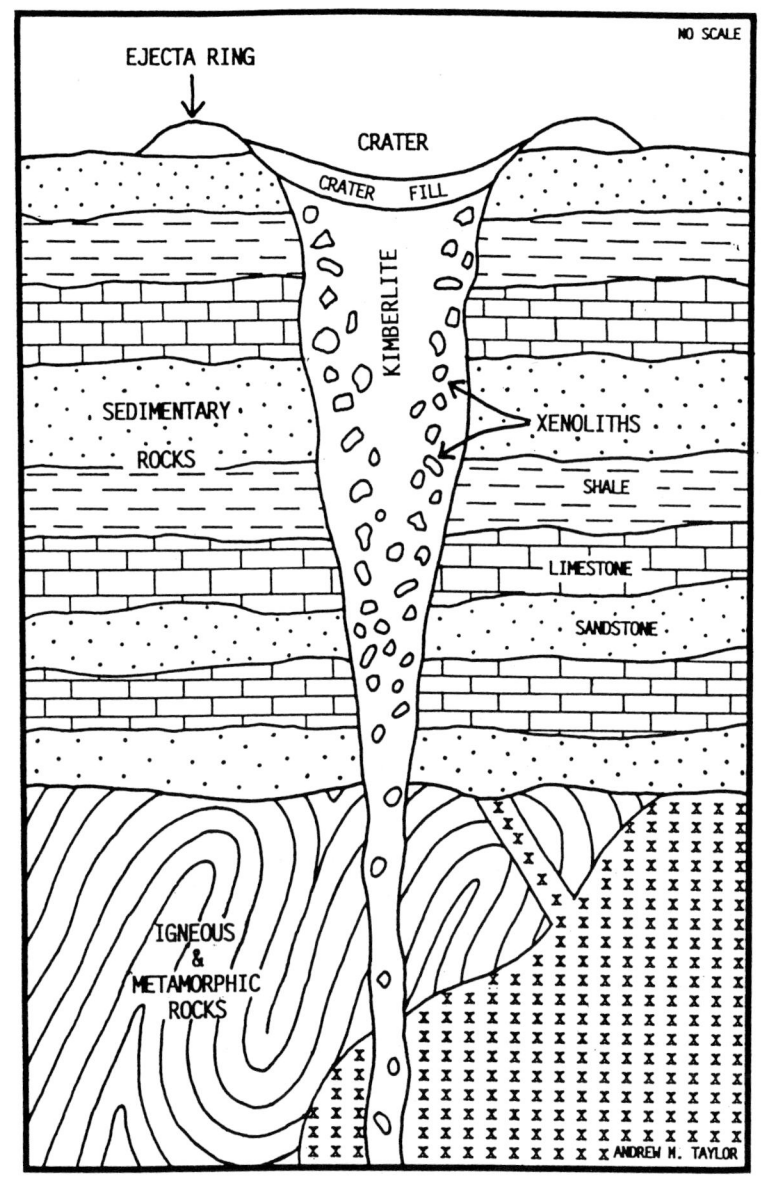

After Figure 19 in Kirkley, Gurney and Levinson, 1991, p. 22.

Figure 15. Kimberlite Diatreme. The geology of "diamond pipes" was first intensively studied in South Africa. This idealized diagram is based on the South African model. According to theory, the kimberlite "magma", containing liquid carbon dioxide under great pressure, moves rapidly up from deep in the mantle. As it nears the surface of the earth, carbon dioxide begins to expand to gas, enlarging the diameter of the "pipe". This creates the carrot-shape of the diatreme. Vaporization of ground water is also thought to contribute to this process. Kimberlites commonly contain xenoliths, pieces of mantle rock and other rocks which the diatreme has penetrated on its passage to the surface. At the surface, the diatreme "blows out", creating a small volcano. Kimberlite sometimes contains diamonds brought up from the mantle.

43

Box 4. Continued

    The "State Line" diatremes of Colorado and Wyoming
contain kimberlites that are largely composed of *serpentine*
(formed by alteration from *olivine*) with abundant fragments
of red *garnet*, green *chrome diopside*, *ilmenite*, and *spinel*.
These *mineral* inclusions are called "indicator" minerals.

    We now turn our attention to the geologists who were
puzzling over the significance of Silurian limestones in an
area where such rocks had no reason to be. From later
discoveries, it was learned the limestone boulders are
actually "*xenoliths*" encased in kimberlites within
diatremes. The youngest rocks among the "xenoliths" are
Silurian limestones. They are age-dated by the fossils
they contain (Chronic, et. al., 1969, p. 151).
Consequently, fossil evidence indicates the diatremes were
emplaced in late Silurian time or early in the Devonian
Period. After the initial eruption, blocks of Silurian
rock broke loose from the diatreme wall and traveled great
distances down the pipe. However, during the eruption,
kimberlitic *magma* moved up through the pipe of the volcano.
Because of density differences, the limestone blocks could
not have sank through the liquid magma, yet somehow they
moved down. Limestones are composed of the *mineral
calcite*, a very light density mineral. Kimberlites contain
heavy elements such as iron. As a liquid, it would still
be more dense than calcite. Limestone blocks would have
floated on the magma as a cork floats on water! It has
been proposed "fluidization cells", a form of a heat
convection cell, may have circulated in the pipe and
carried xenoliths of wall rock downward. Another possible
explanation is the volcanic pipe was essentially empty for
a period of time and zenoliths simply fell down the pipe.
Others have suggested it was filled with a very frothy
liquid magma (containing much gas) which would allow the
low-density limestone to sink down within the pipe.

    Erosion during the Devonian Period later removed the
overlying Cambrian, Ordovician and Silurian rocks. So,
much of the diatremes have been eroded away and what is
left is surrounded by very old Precambrian *granites* and
metamorphic rocks (See Figure 16).

    In summary, this sequence of geologic discoveries
rewrote geologic history. The Silurian limestones in
diatremes is proof Colorado and Wyoming were covered by
extensive Silurian seas. These Silurian rocks which were
once present were removed by erosion. Additionally, it was
learned diamond pipes do exist in Colorado.

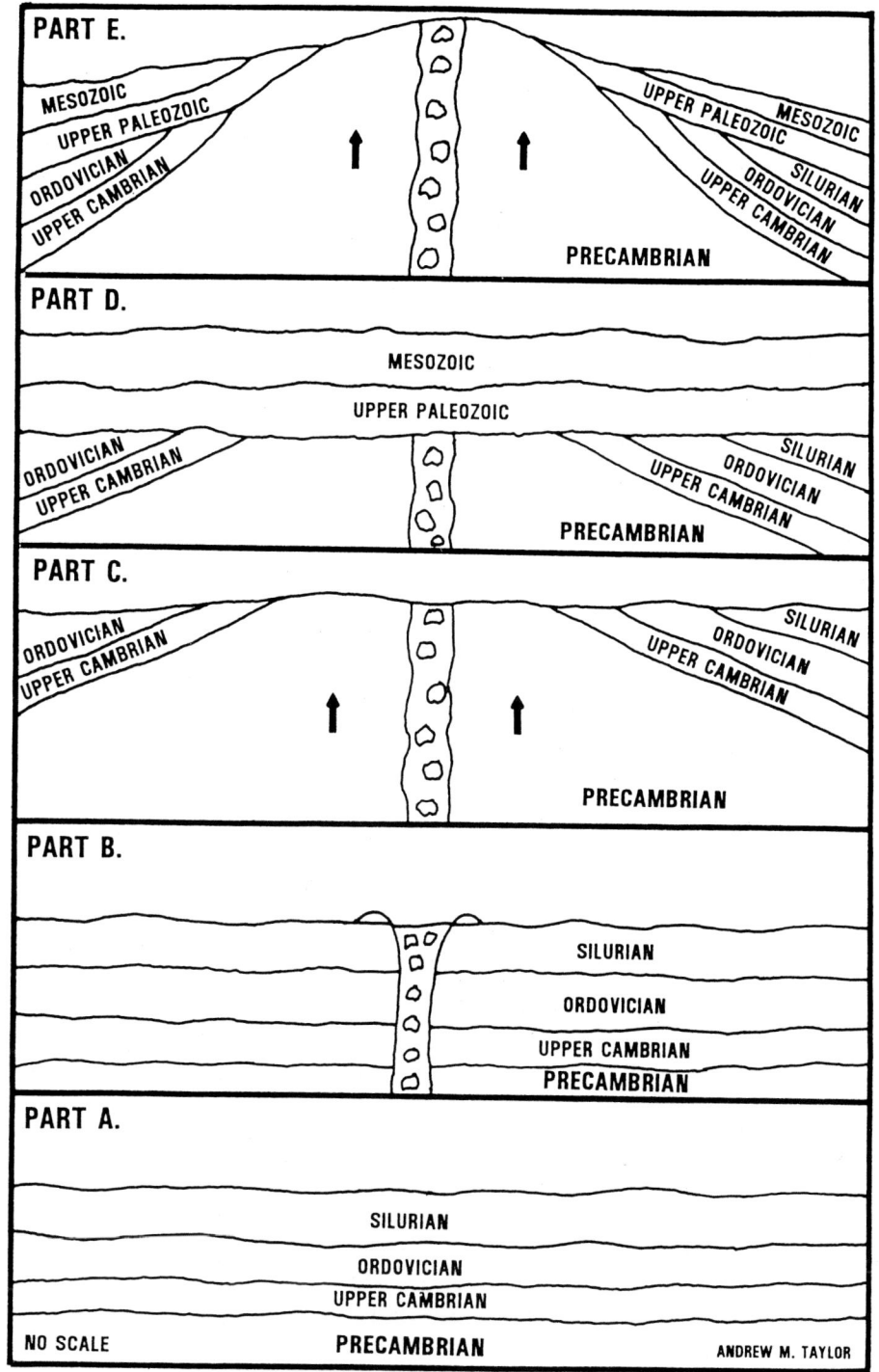

After Chronic, 1976, p. 105 and Tweto, 1980a, p. 7.

Figure 16. Colorado Diamond Pipe. An idealized diagram of a typical Colorado diatreme. The roots of the diatreme extend deep into the mantle.

Further discussion of Figure 16 is continued on the next page.

45

The time of the intrusion of the State Line District diatremes is problematic. Blocks of Ordovician and Silurian limestones found within the diatremes are age-dated by fossils. These demonstrate that the diatremes are younger than Silurian, but how much is not known. Some non-fossiliferous blocks are lithologically similar to Devonian and Mississippian limestones in other areas (Chronic et. al., 1969, p. 154). It is safe to assume that the diatremes are of Early Devonian age and possibly younger (Late Devonian or Mississippian).

Part A. Upper Cambrian, Ordovician and Silurian rocks (mostly marine) were deposited on top of Precambrian rocks over Colorado.

Part B. Intrusion of the diatremes during Early Devonian.

Part C. Middle Devonian uplift and erosion. The Lower Paleozoic rocks were eroded back and Precambrian rocks exposed.

Part D. Upper Paleozoic and Mesozoic rocks were deposited. The State Line District was located off the flank of the Ancestral Rockies and was not uplifted at that time (300 million years ago).

Part E. Uplift of the modern-day Rocky Mountains. The Upper Paleozoic and Mesozoic rocks were eroded back from the summits, leaving Precambrian rocks with diatremes exposed at the surface.

Box 4. Continued

DIAMOND MINING IN COLORADO

At least two attempts have been made to commercially mine the Sloan Diatreme in the State Line District. Although more than 100,000 diamonds were recovered, both efforts were unsuccessful. Most diamonds were quite small and only a small percentage were of gem quality. There simply were not enough diamonds in the kimberlite to be mined at a profit.

The Kelsey Lake Diatreme (also in the State Line District) appears to be a different story. Redaurum, a Toronto-based mining company, started a commercial operation at the Kelsey Lake kimberlite deposit in 1995. It

---

Box 4. Continued

appears to have been a successful venture until late 1997, when it was closed. Kelsey Lake was the only commercial diamond mine operating in the United States in recent times.

In late 1996, the mine recovered a gem-quality 28 carat diamond which has a yellowish tint. This is the fifth largest diamond ever found in North America! The 28 carat diamond was cut into a 5.39 carat deep yellow, pear-shaped stone. It was sold for $87,000 to an undisclosed buyer. In late 1997, another large diamond was announced. After cutting, it weighed 16.68 carats and is the largest cut diamond ever found in the United States. The original weight of the stone before cutting was reported to be 28.20 carats. The first cut diamonds from the Kelsey Lake Mine were, for a period of time, sold at Hyde Park Jewelers in Denver. Coloradans and visitors to the state had the opportunity to buy real Colorado diamonds.

Litigation over mineral rights at the Kelsey Lake mine began in 1997. The mine is currently for sale.

So, 120 years after the embarrassment of the "Great Diamond Hoax", Colorado really does have diamonds!

DEVONIAN PERIOD: 410-360 *M.Y.A.*

EARLY DEVONIAN:

There are no Early Devonian-age rocks present in Colorado. It is assumed that Colorado was *emergent* and no rocks were deposited or preserved until the Late Devonian. An unconformity thereby exists below the rocks of the Late Devonian (See Figure 17).

LATE DEVONIAN:

Colorado was again covered by a tropical ocean and *limestones* were deposited.

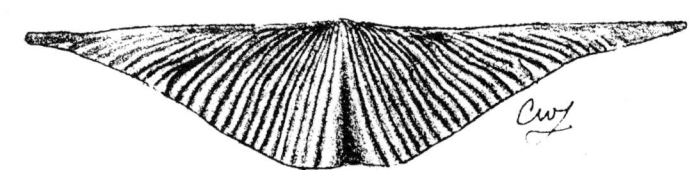

Devonian Brachiopod (Spirifer)

Chaffee Group: Consists of a lower formation known as the Parting Sandstone which contains *fossil* fish

47

and an upper formation, the Dyer Dolomite. Various *shellfish* and *coral* fossils are found in the Dyer. Near the middle, the Dyer also contains thin-bedded *chert* and *chert nodules*.

Figure 17. Unconformity. This photo, taken in Box Canyon near Ouray, demonstrates an obvious unconformity in the rocks. Horizontal beds of the Devonian Elbert Formation (equivalent to the lower Chaffee Group) lie unconformably on top of the Precambrian Uncompahgre Formation (vertical beds).

MISSISSIPPIAN PERIOD: 360-330 *M.Y.A.*

Colorado was covered by a warm shallow ocean that contained many sea creatures. The abundance of *fossiliferous* limestones deposited on the sea floor tells us life was thriving on earth during this time.

Leadville Formation: Gray massive *limestone* or *dolomite* containing *shellfish* and *coral* fossils. This formation is often referred to as the Leadville Limestone, but in Colorado, it is commonly a dolomite. The Leadville also contains dark to black *chert*. In Western Colorado, part of the formation contains *oolitic limestones*.

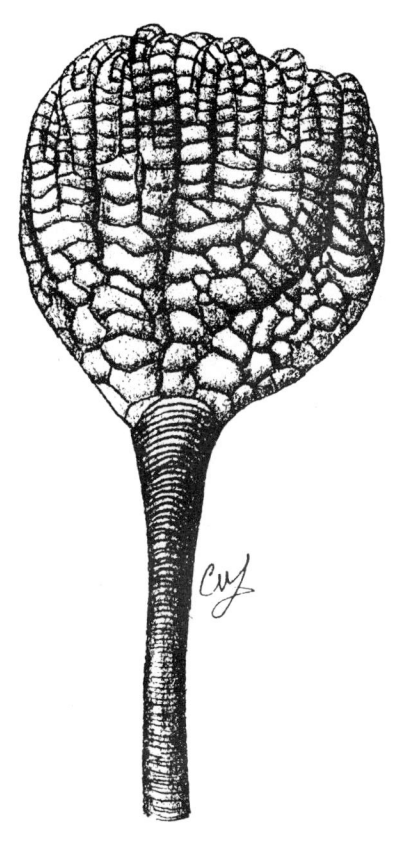

Mississippian Crinoid

PENNSYLVANIAN PERIOD: 330-290 *M.Y.A.*

Colorado began to rise, and plate collisions occurred between North American and the continents of Africa and Europe. The merging of these continents plus Asia created a giant supercontinent known as *Pangaea*. Crustal compression from the collision formed the Appalachian-Ouachita *fold* belt.

Many vertical uplifts in Colorado, Wyoming, New Mexico, Oklahoma and Texas occurred as a consequence of compression from plate collision. Two of these uplifts were in Colorado: the Ancestral Rockies which trended northwest through central Colorado and the Uncompaghre Uplift in southwestern Colorado (See Figure 18). These were literally mountains in the sea (which covered the rest of Colorado). *Weathering* and *erosion* of these mountains created extensive *alluvial fans* at the interface of the mountains and the ocean. Ultimately, these uplifts eroded away and no longer exist. Box 5 presents the geological evidence for the Ancestral Rocky Mountains.

During Pennsylvanian time, the ocean basin in central Colorado (between the Ancestral Rocky Mountains and the Uncompaghre Uplift) became "*restricted*" (separated or blocked from the open ocean) at times. As ocean water evaporated, thick deposits of *gypsum* and salts were precipitated. These *evaporites* are part of the Minturn and Maroon formations. Gypsum deposits may be seen along Interstate I-70 near the town of Gypsum.

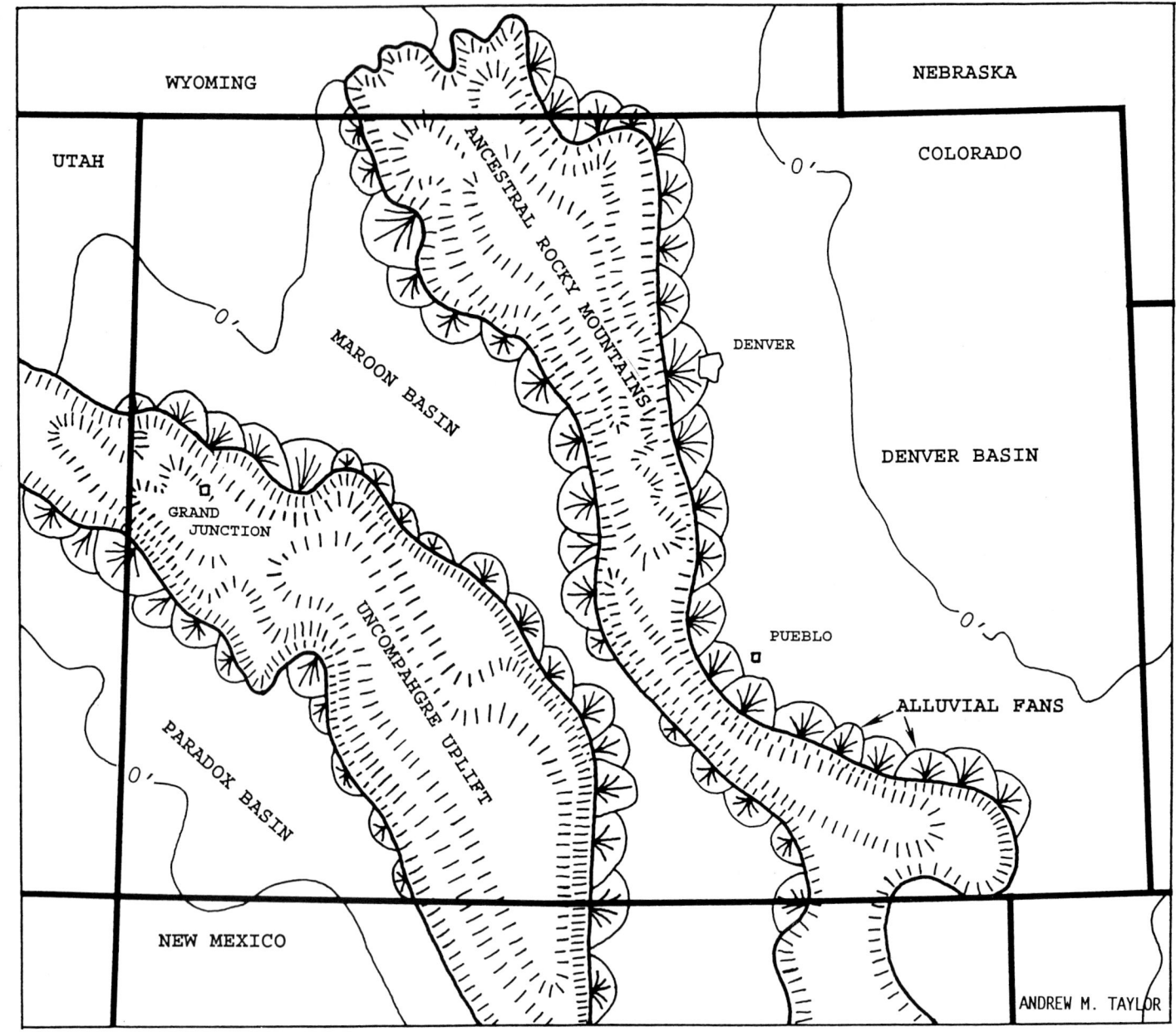

After Mallory, 1958 and 1972.

Figure 18.   Pennsylvanian Ancestral Uplifts of Colorado.
During the Middle Pennsylvanian, the Ancestral Rocky Mountains
and the Uncompahgre Uplift rose out of the ocean creating
mountain ranges surrounded by the ocean. Large alluvial fans
formed at the interface of the mountains and the ocean. On
the eastern side of the Ancestral Rocky Mountains, these fans
form the Fountain Formation. Between the Ancestral Rockies
and the Uncompahgre Uplift, the fans are known as the Minturn
and Maroon formations. On the western side of the Uncompahgre
Uplift, they are known as the Hermosa Formation. Thick
sequences of evaporites (salt and gypsum) are included in the
Minturn and the Hermosa. At times, the ocean basins in the
areas where these formations were being deposited were cut off
from the rest of the ocean. Large amounts of marine water
evaporated, leaving behind thick deposits of evaporites.

50

On the southwest side of the Uncompaghre Uplift, the ocean basin (referred to by geologists as the Paradox Basin) also became restricted at times and thousands of feet of evaporites (gypsum, salt and potash) were deposited. These evaporites are interbedded with rocks of the Hermosa Group. Under the weight of overlying sediments, salt beds in the Hermosa flowed and formed several salt *anticlines*. On today's landscape, the anticlines are located southwest of the Uncompaghre Plateau. Growth of the anticlines by salt flowage began in the late Pennsylvanian and continued into the Tertiary Period. In late Tertiary and Pleistocene time, ground water dissolved much of the salt and the anticlines collapsed inward, forming "breached" anticlines.

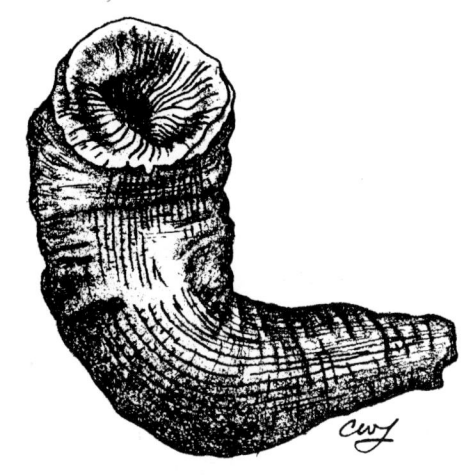

**Pennsylvanian
Horn Coral**

EASTERN COLORADO:

Glen Eyrie Formation: Similar to the Belden Formation of central Colorado with *marine shale*, *sandstone* and *limestone*.

Fountain Formation: The Fountain unconformably overlies Precambrian rocks (See Figure 19). It is primarily composed of red *cross-bedded* conglomeratic *feldspar*-rich *sandstones* and lesser amounts of thin dark red *mudstones* which were deposited in a *nonmarine alluvial fan* environment (See Figure 20). The Fountain may be up to 1,000 feet thick.

BOX 5.   GEOLOGICAL EVIDENCE FOR
THE ANCESTRAL ROCKY MOUNTAINS

The Ancestral Rocky Mountains no longer exist. They slowly eroded away and disappeared into the mist of geologic time. All we have left of the ancestral mountains are certain lines of geological evidence which indicate they did exist. You might say they left their "footprints" on the Pennsylvanian landscape. Similar lines of evidence establish the existence of the Uncompahgre Uplift in the southwestern portion of the state. An idealized diagram

---

51

Box 5. Continued

showing the uplift and subsequent erosion and burial of the ancestral uplifts is presented in Figure 21. An overview of this evidence follows:

1. **PENNSYLVANIAN-PRECAMBRIAN UNCONFORMITY:**

    The Pennsylvanian Fountain Formation overlies the rocks of the much older Precambrian Idaho Springs Formation. The contact between them is an erosional surface, an *unconformity*, representing a time gap of a minimum of 270 million years (See Figure 21). The missing rocks include those of the Cambrian, Ordovician, Silurian, Devonian, Mississippian and Lower Pennsylvanian ages, with an aggregate thickness from 6,000 to 12,000 feet (one to two miles!). Clearly, this is a major unconformity created by *weathering* and *erosion* of a great amount of rock. Let us now examine the geologic requirements necessary to erode away so much rock.

    Initially, weathering and erosion occur at the surface of the earth. In order to erode away one to two miles of rock, they have to be continually uplifted as they erode. If rocks are uplifted by one to two miles, the geologic process involved is probably mountain-building! This major unconformity is good evidence of an old mountain range that existed at this time. Geologists have named it the "Ancestral Rocky Mountains". This old mountain range slowly eroded away as millions of years passed until it disappeared, never to be seen again except in the minds of geologists! It must be remembered that the Ancestral Rocky Mountains have nothing to do with the modern-day Rocky Mountains.

2. *Provenance of the Fountain Formation:*

    The Fountain Formation is believed to be a series of coalescing *alluvial fans* deposited on the eastern flank of the eroding Ancestral Rocky Mountains, as shown in Figure 18. The fans are composed of *rocks* and *minerals* eroded from the mountains. If this is true, then rocks of the Fountain should contain rock and mineral types compatible with those of the rocks of the Ancestral Rocky Mountains. Examination of the minerals and rocks of the Fountain reveals they are *quartz* and *feldspar* minerals and pebbles, cobbles and boulders of *gneiss*, all of which are compatible with those of the Ancestral Rockies. On the western side

Figure 19. Fountain/Precambrian Unconformity.
Sandstones and conglomerates of the Fountain Formation
are located on the left side of the photo and gneisses of
the Precambrian Idaho Springs Formation are on the right.
The contact between the two formations runs diagonally
from the lower left to the upper right. This is a major
unconformity and rocks of Cambrian through Lower
Pennsylvanian age are missing. The missing rocks
represent a thickness of approximately one to two miles.
This major unconformity is regarded as evidence for the
Ancestral Rocky Mountains. The missing rocks are
believed to have eroded away from their summits as the
mountains uplifted. By mapping the unconformity,
geologists are thereby able to map the geographic extent
of the Ancestral mountains.

Box 5. Continued

of the Ancestral Rockies, the equivalent alluvial fans are known as the Minturn Formation and the Maroon Formation (lower part).

3. **LARGE SIZE OF CLASTICS NEAR SOURCE:**

The Fountain Formation contains *conglomerate* lenses composed of pebbles, cobbles and boulders. Large size *clastics* such as these are deposited near the source from which they eroded. This supports the concept that the Fountain sediments were derived from the nearby crests of the eroding Ancestral Rocky Mountains. The highest parts of the mountains were some fifteen to twenty miles west of present-day exposures of the Fountain.

4. **PRESENCE OF FELDSPAR INDICATES NEARBY SOURCE:**

Feldspar is an unstable mineral and under weathering conditions, rapidly decomposes to *clays*. Therefore, feldspar will not travel far from the erosional source before it decomposes. The Fountain contains abundant feldspar pebbles and grains. This implies that the erosional source of the feldspar was nearby, i.e., the Ancestral Rocky Mountains.

5. **GEOMETRY OF THE ALLUVIAL FANS OF THE FOUNTAIN FORMATION:**

Alluvial fans have a fan-shaped geometry and thin rapidly toward their edges. In Colorado, there are many modern-day alluvial fans currently being constructed by streams draining from mountain areas. They may be easily seen if you look for them. In 1966, a geologist by name of Howard mapped nine large paleoalluvial fans in the surface and subsurface of the rocks of the Fountain Formation. These ancient fans extend from southern Wyoming down along the leading edge of the Front Range to the border with New Mexico. In the subsurface to the east of the Front Range, the fans rapidly thin and ultimately interfinger with *marine rocks* under eastern Colorado. For example, the Fountain alluvial fans which are exposed in the Red Rocks Park area, thin to the east in the subsurface and are essentially gone under the city of Denver.

Figure 20. Fountain Formation. Large "flat irons" composed of red sandstones of the Fountain Formation are spectacularly exposed in the northern end of Roxborough Park, as well as in other areas along the eastern edge of the Front Range. The Fountain Formation has an average thickness of 1,000 feet. It is largely composed of red coarse-grained sandstone with some conglomerate and thin dark red mudstones. Where the sandstones are well-cemented and resistant to erosion, they are exposed as "flat irons." At the time these layers of sandstone were being deposited, they were near horizontal. The uplift of the modern-day Rocky Mountains to the west tilted these beds toward the east.

55

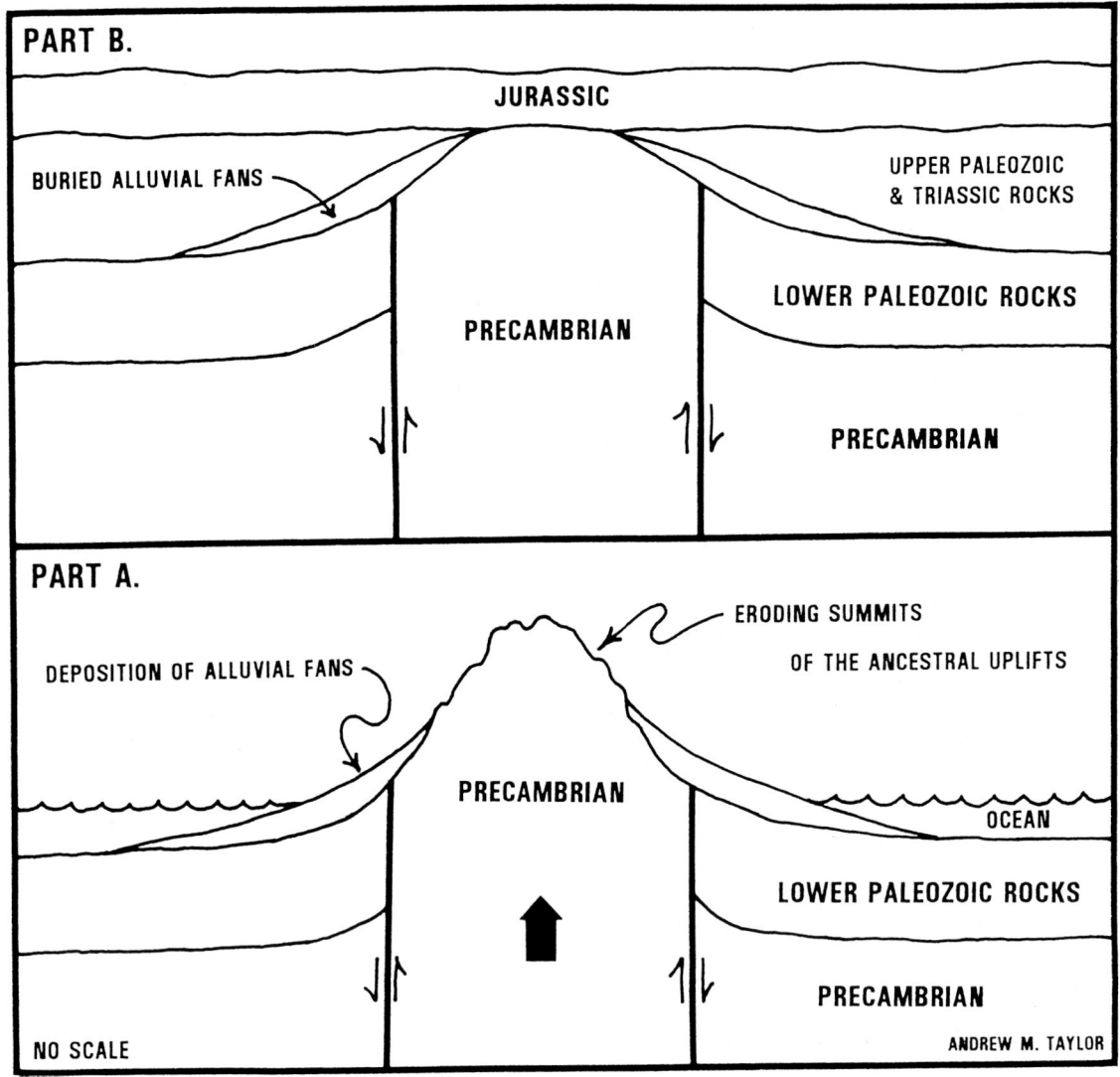

Figure 21. Ancestral Uplifts. The uplift and subsequent erosion and burial of an ancestral uplift is shown in this idealized diagram. In Part A, the uplift of a block of the earth's crust forms an ancestral mountain range which begins to erode with alluvial fans being deposited on its flanks. In Part B, the mountains have eroded away and have been covered by younger Jurassic sediments. The alluvial fan deposits contain conglomerate and feldspar grains derived from the Precambrian rocks.

56

CENTRAL COLORADO:

Sediments were deposited between the Ancestral Rocky Mountains and the Uncompaghre Uplift in what is known as Maroon Basin (sometimes termed the Central Colorado Basin). *Arkosic conglomerates* in the these rocks record the initial uplifting of the Ancestral Rocky Mountains and the Uncompahgre Uplift.

Belden Formation: Dark gray *shale*, *sandstone* and *fossiliferous limestone*. This formation was deposited as Colorado was rising and the sea was retreating and shallowing. At this time, there were no mountains yet.

Minturn Formation: The Minturn was created by cyclic deposition of red and gray *shale*, *sandstone* and *conglomerate*, with some *limestone* beds and *evaporites*. It coarsens upward (increasing amounts of sandstone and conglomerate in the upper part of the formation) reflecting the beginning of the rise of the Ancestral Rocky Mountains and the Uncompahgre Uplift. The Minturn is equivalent to the lower portion of the Fountain Formation of eastern Colorado.

Maroon Formation: Bright red *shale*, *siltstone*, *arkose* and *conglomerate* derived from the *erosion* of the Ancestral uplifts and deposited on *nonmarine floodplains* or "mudflats". Also contains *evaporites*. The Maroon is equivalent to the upper portion of the Fountain Formation of eastern Colorado. At the time, the central portion of Colorado must have been extremely unattractive.

WESTERN COLORADO:

Hermosa Group: This group has complex *stratigraphy* and is composed of a number of formations, some of which are further subdivided into members and "zones". Largely composed of *sandstones*, *shales* and *evaporites*, the Hermosa Group is similar to the Minturn and Maroon Formations of Central Colorado. The Hermosa was derived from *erosion* of the Uncompahgre Uplift and deposited on the west flank in the Paradox Basin.

The Fountain, Minturn, Maroon and Hermosa formations were all derived from erosion of Pennsylvanian-age uplifts and therefore have a common origin.

PERMIAN PERIOD: 290-240 *M.Y.A.*

By this time, the Ancestral Rocky Mountains and the Uncompahgre Uplift were eroded down to hills. Streams now flowed with slower velocities and they generally carried nothing larger than sand, silt and clay. Occasional flood stages caused small pebbles to be carried and deposited. The type of *rock* and *mineral* grains (pebbles of *gneiss* and fragments of *feldspar*) in these *sandstones* and *conglomerates* is still compatible with the presumed source from the hills of the Ancestral Rocky Mountains and the Uncompahgre Uplift.

EASTERN COLORADO:

Lyons Formation: Red, pink, and white *sandstones* deposited by *braided streams* and also as *eolian* dune sands. The ocean was still nearby to the east. Near the city of Lyons, the sandstones are thinly layered in sheets. They are quarried and used all over the Denver metropolitan area as flagstone and building stone (See Figure 22). Frequently, the surfaces (bedding planes) of the flagstones show *paleoraindrop impressions* and the Lyons is famous (at least among geologists!) for these (See Figure 23). The sandstones here have been variously interpreted by different geologists as *beach* deposits at the edge of the ocean or as eolian sand dunes. A number of features in these sandstones indicate they were formed in a large eolian sand dune field in this area. The paleoraindrop impressions can only form on dry sand. Slump

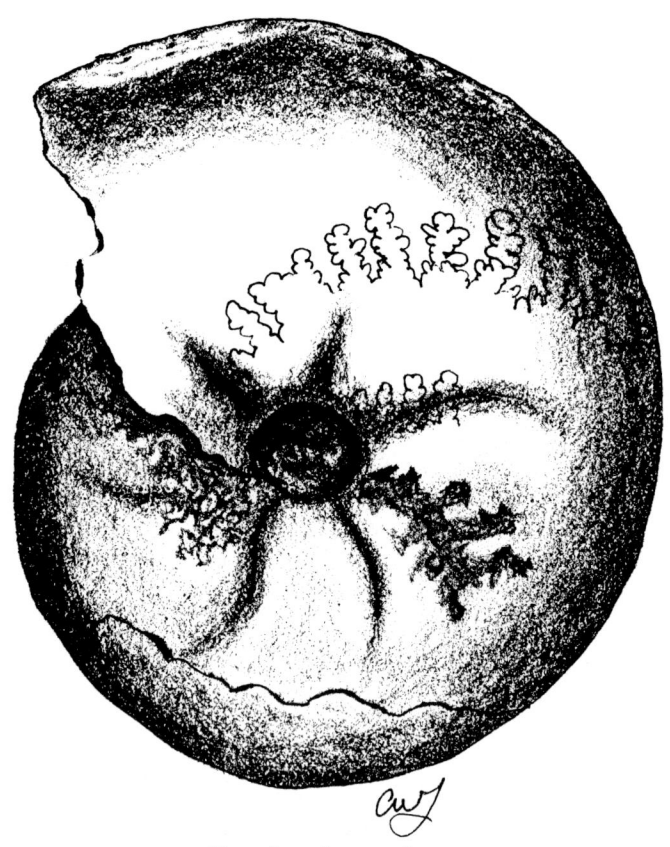

Permian Ammonite

features (very small "landslides") have been observed, similar to slumps on the *slip faces* of modern sand dunes. Animal tracks believed to have been made by reptiles have been found on bedding planes of the sandstones, all of which infer a *subaerial* origin (Walker and Harms, 1972, p. 279-288).

At the town of Morrison, the lower Lyons consists of white *cross-bedded* sandstones deposited in river channels and as windblown sand dunes in the upper part of the formation. The bedding planes of the river channel sandstones frequently exhibit *paleomudcracks* (See Figure 24). When wet sand and mud dries out, it shrinks and cracks develop. Mudcracks are very common in modern arid environments (See Figure 25). The abundance of paleomudcracks, paleoraindrop impressions and eolian sand dunes in the Lyons Sandstone suggests an arid environment. At this point in geologic history, the North American plate was moving north, out of the tropics and into the great desert belt lying between 30° and 40° north latitude. Colorado was beginning its great desert phase.

Figure 22. Lyons Sandstone.   Sandstones in a quarry at Lyons, CO. In this area, the Lyons Sandstone is composed of thin, parallel layers of sandstone. These have been mined and used extensively as flagstone and building stone along the Front Range. The sandstone is believed to have been deposited as cross-strata in the slip faces of sand dunes. An extensive sand dune field is presumed to have existed in this area during the Late Permian. Photo by W.T. Lee, 1921. USGS Photo Library.

Figure 23. Paleoraindrop Impressions. The circular features on this specimen of Lyons sandstone are raindrops which fell on the sand before it was lithified into hard sandstone. The preservation of something as ephemeral as a raindrop impression requires the following circumstance and sequence of events. There must be dry sand present. A few raindrops fall and create "splatter cones" on the dry sand surface. If it continues to rain, they will all be washed away. After the "splatter cones" are formed, they must be rapidly covered by windblown sand. Only then may they be preserved in the resulting rock. Many bedding plane surfaces in Lyons flagstone quarries exhibit these paleoraindrop impressions. The photograph was taken in the old Lyons flagstone quarry which is now within Boulder Mountain Park.

60

Figure 24. Paleomudcracks. Ancient mudcracks are visible on the base of a sandstone bed in the Permian Lyons Formation located on the western side of the village of Morrison. Sand later filled the original cracks creating a polygonal pattern. Paleomudcracks are abundant on bedding planes of numerous sandstone beds at this locality. These are evidence of an arid environment.

Figure 25. Modern Mudcracks. Water is part of the volume of wet mud and sand. When the mud or sand dries, the water evaporates. The volume of the sediment is reduced and therefore shrinks. Cracks form to accommodate the shrinkage. These cracks have a polygonal pattern and extend downward a few inches at most. Windblown sand may later fill the cracks. This photo shows recently formed mudcracks in what is obviously an arid environment. Photo by G.K. Gilbert, circa 1907. USGS Photo Library.

WESTERN COLORADO:

The Maroon Formation continued to be deposited.

Weber Formation: The Weber is a thick *sandstone* present in far
northwestern Colorado. It is stratigraphically equivalent to
lower parts of the Maroon Formation and represents *eolian*
sandstone deposits. The sand in the Weber is believed to have
come from sources in Wyoming. It outcrops on the flanks of
the Uinta Mountains and is particularly well displayed in
Dinosaur National Monument. It forms a major portion of the
magnificent canyon walls in the vicinity of the confluence of
the Green and Yampa rivers. In early Permian time, there was
a major field of sand dunes in this part of Colorado and to
the southeast, the red *mudstones* of the Maroon were being
deposited.

LATE PERMIAN:

The Ancestral Rockies and the Uncompahgre Uplift were eroded
down to very low hills. Sluggish streams which drained from the
hills meandered through the *floodplains*. The ocean was still
present in eastern Colorado and these floodplains were not much
above sea-level. In fact, some may have been within the inter-
tidal range. Streams had low velocities and were only able to
carry smaller grain sizes (silt and clay). Extensive red mudflats
which were probably dry most of the time flanked the low hills over
much of Colorado. It is difficult to think of Colorado as ever
being ugly, but it probably was during this time! Who likes
mudflats?

EASTERN COLORADO:

Lykins Formation: Composed of red, *micaceous mudstones* (a mixture
of silt and clay). The ocean was nearby to the east and the
Lykins may have at times been within the tidal zone. Red
mudstones were deposited on arid *floodplains* during flooding.
Twice the ocean transgressed over the land and deposited the
thin *marine limestone* members of the Lykins, the Falcon and
the Glennon (See Figure 26). After deposition of the Lykins,
the Ancestral Rockies were completely eroded away and their
"roots" were buried by younger sediments.

WESTERN COLORADO:

The Maroon Formation continued to be deposited.

MESOZOIC ERA

TRIASSIC PERIOD: 240-205 *M.Y.A.*

The first dinosaurs appeared and were the dominate land life forms of the Mesozoic. During the Triassic, the North American Continent was drifting north and moving through the northern desert belt. Colorado was generally a low-land area with an arid climate. The landscape continued to be dominated by somewhat less than scenic red mudflats.

EASTERN AND CENTRAL COLORADO:

The Lykins Formation continued to be deposited.

Figure 26. Permian/Triassic Lykins Formation. On the west side of Morrison, red mudstones and a thin white limestone of the lower Lykins Formation are exposed. Red mudstones were deposited on arid, nonmarine floodplains. Limestone was deposited during a brief marine transgression from the east.

WESTERN COLORADO:

Rocks similar to the Lykins continued to be deposited, but have many different names.

Moenkopi Formation: Red to brown *shale* and *sandstone* deposited on tidal flats.

Chinle Formation: Red *shale*, *siltstone* and *mudstone* with some *sandstone* and *conglomerate*. May contain teeth and bones of reptiles.

Glen Canyon Group: The Glen Canyon Group is composed of the Wingate and Kayenta formations and the Navajo Sandstone. The age of the Glen Canyon Group has been the subject of much controversy. These are arid deposits which are notoriously non-*fossiliferous* and their ages can only be determined by correlation with units in other areas. This is not always an exact science. Some geologists believe it may be of lower Jurassic age. Many petroleum geologists regard the Navajo as being of Jurassic age. In conformance with the U.S. Geological Survey, the author has assigned the lower Glen Canyon to the Triassic and the upper (Navajo) to the Jurassic, with a question mark.

Wingate Formation: Thickly *cross-bedded sandstone* derived from extensive sand dunes. Forms massive cliffs and canyon walls.

Kayenta Formation: Buff, gray, red and purple irregularly bedded *shale*, *siltstone*, *sandstone*, *conglomerate* and *limestone*.

JURASSIC PERIOD: 205-138 *M.Y.A.*

EARLY JURASSIC:

During this time, Colorado was flat and was probably a dry desert.

EASTERN COLORADO:

No Early Jurassic rocks are present in Eastern Colorado.

WESTERN COLORADO:

Early Jurassic rocks are present and bear many formational names. The ones presented in this publication have widespread usage.

> **Navajo Sandstone:** *Cross-bedded sandstones.* An extensive field of wind-blown sand dunes covered much of Western Colorado. This area may have resembled the Sahara Desert of today. The exact age of the Navajo Sandstone is problematic. Many geologists regard it as being Jurassic.

> **Carmel Formation:** Red *shale*, *siltstone* and *mudstone*. Sometimes *gypsiferous*. More red mudflats!

> **Entrada Sandstone:** *Cross-bedded sandstones.* An extensive field of wind-blown sand dunes formed. Once again, western Colorado must have resembled the Sahara Desert.

LATE JURASSIC:

EASTERN COLORADO:

Ralston Creek Formation: *Sandstones*, *shales* and *gypsum evaporites* deposited on a *nonmarine floodplain*. The formation is similar in nature and mode of origin to the Summerville Formation of western Colorado. Eastern Colorado was still near sea-level and the gypsum evaporites are believed to be *lagoonal* deposits. The Ralston Creek rocks are soft. They weather and erode easily. As a consequence, it is a "*valley former*" and is usually covered by soils.

Morrison Formation: *Nonmarine claystones* with lesser amounts of *sandstone* and *limestones*. The claystones originated on *floodplains* and are colored gray-green and maroon. These colors are diagnostic of the Morrison. Sandstones were deposited in rivers and lakes. Limestones are freshwater lake deposits (See Figure 27). The relative abundance of the lakes indicates the landscape of the Morrison was quite flat and poorly drained.

When the Morrison Formation was created, Colorado was a coastal plain with relatively uniform depositional conditions across the entire state. As such, the Morrison is the only formation deposited across the entire state. It is easily recognizable wherever you encounter it.

The Morrison Formation contains a wealth of petrified dinosaur bones. These have been found in many areas in the state where rocks of the Morrison are exposed. On public lands, these may only be collected by permit. For a discussion of dinosaurs, see Box 6.

By this time, the climate had changed and now was humid with abundant growth of trees and plants on the floodplains. The numerous large dinosaurs present on the Morrison landscape needed considerable vegetation to feed on.

Figure 27.   Limestones of the Morrison Formation. Several thin resistant beds of limestone are exposed in the Jurassic Morrison Formation in the Interstate 70 roadcut west of Denver. They contain fossils of calcareous algae known to have lived in fresh water lakes. The abundance of these thin, discontinuous lake deposits indicates the Morrison landscape was relatively flat and poorly drained.

BOX 6. THE DINOSAURS

The dinosaurs ruled the earth for about 140 million years. They evolved in the Triassic Period, and, in some respects, reached a peak of evolution in the Jurassic Period. They became extinct at the end of the Cretaceous Period. The largest number of dinosaurs species existed during the Jurassic. It was also during the Jurassic when dinosaurs reached their largest size. Later, in the Cretaceous, the dinosaurs were generally smaller in size. There were fewer species, but some species had enormous numbers of individuals.

Colorado has extensive exposures of the dinosaur bone-bearing Jurassic Morrison Formation, much of which is located on public lands. The reader should be aware that collection of dinosaur bones on public lands is prohibited by law. Take a bone and go to jail!

WHAT IS A DINOSAUR?

Early *fossil* discoveries were regarded as dead members of presently existing species. This was in accordance with religious beliefs of the time. In 1770, jaws of a huge *marine* lizard (Mosasaurus) were found in Holland and recognized by the eminent French anatomist Baron Georges Cuvier as being an extinct species. Subsequently, many other fossil "reptiles" were found.

Sir Richard Owen (1841), reviewed British fossil reptiles and discovered some were different from others. The legs of reptiles splay out sideways from the body as befits a crawling mode of transportation. Owen observed that some of the fossilized "reptiles" had legs tucked under the body (as in mammals) and were therefore a different species. He named them "dinosaurs" ("terrible lizards").

COLORADO DINOSAURS:

EASTERN COLORADO:

The first great dinosaur discoveries in the U.S. occurred in Colorado almost simultaneously in the quarries of Garden Park near Canyon City and in the hogback near Morrison. As reported in the local newspaper in January of 1877, a high school teacher named Ormel W. Lucas discovered dinosaur bones at Garden Park. In March, Lucas wrote a letter describing his discovery to Edward Drinker Cope of

Box 6. Continued

the Philadelphia Academy of Sciences. Also in March, Arthur Lakes, a high school science teacher (and later a Professor at the Colorado School of Mines) discovered large dinosaur bones outside Morrison at what is now called Dinosaur Ridge (See Figure 28). Lakes immediately contacted Othniel Marsh of Yale Peabody Museum to inform him of the discovery and even shipped boxes of bones to him. Marsh ignored him until he learned Lakes also sent bone specimens to Marsh's chief rival, Edward Drinker Cope. As a consequence of these actions, Marsh controlled the quarries at Morrison and Cope the quarries at Garden Park (See Figure 29), although Marsh also later opened a quarry at Garden Park (See Figure 30).

Figure 28. Dinosaur Bone. In the center of the sandstone boulder, a portion of a dinosaur leg bone is exposed. This may be viewed at Dinosaur Ridge near Morrison.

69

Figure 29. Cope Dinosaur Quarry, Garden Park. Site of the 1877 Cope and Hatcher dinosaur quarries one mile west of Felch's Ranch. A quarry and associated dumps are visible in the lower left hand corner of the picture. Photo by T.W. Stanton circa 1910. USGS Photo Library.

Figure 30. Marsh Dinosaur Quarry, Garden Park. Site of the 1877 Marsh dinosaur quarry on Felch's Ranch. The quarry is located above the sandstone ledge. Dump material was shoved over the ledge, creating a dump below. The dump is visible in the right center of the photograph. Photo by T.W. Stanton, circa 1910. USGS Photo Library.

Box 6. Continued

Marsh and Cope became bitter rivals in the collection, description and naming of dinosaurs. Their struggles are known as the "Marsh-Cope Wars." It has been said their crews even fought over dinosaur bones! From 1877 to the late 1890s, Marsh and Cope named 130 new species of dinosaurs, a veritable avalanche!

The rivalry intensified in 1870 when Marsh pointed out Cope had mounted the head of a large marine "lizard" on the wrong end of the skeleton he had reconstructed! Cope never forgave Marsh for publicizing his mistake. Later, Cope got his revenge when it became known Marsh had placed the wrong head on a dinosaur reconstruction.

The competition between Marsh and Cope resulted in the collection of enormous numbers of dinosaur bones to be collected. Even today, most of them have yet to be prepared and described. Such work is very time consuming.

In the latter part of the Nineteenth Century, Marsh and Cope were giants in the scientific world of the United States. In those days, a scientist had to join one or the other camps in order to obtain employment. Such was their influence.

Because of the haste to find, name, and publish new species of dinosaurs, mistakes were made. Marsh gave the name, Titanosaurus montanus, to a new find from the Morrison area. A short time later, he changed the name to Atlantosaurus when he learned the name Titanosaurus had already been used for a dinosaur from India. Still later, Marsh gave the name Apatosaurus to some additional fossils found by Arthur Lakes at Morrison. He also described two new species of dinosaurs, Brontosaurus and Morosaurus from a find at Como Bluff, Wyoming. Meanwhile Cope used the name Camarasaurus for a dinosaur he collected at Garden Park. In later years, when the feverish pace of the rivalry had died down, other scientists carefully reexamined these dinosaurs and decided these names were applied to only two species of dinosaurs. Consequently, Apatosaurus and Camarasaurus were the surviving names as follows:

Hence: Titanosaurus ⌉
           Atlantosaurus = Apatosaurus
           Brontosaurus ⌋

                   and
           Morosaurus = Camarasaurus

Box 6. Continued

Today, the name Brontosaur is strictly an informal name applied to large Jurassic herbivorous dinosaurs.

It is difficult to unravel the history of the new dinosaur species discovered and named by Marsh and Cope in their feverish rivalries at Dinosaur Ridge and Garden Park. Marsh, for example, sometimes announced a new dinosaur species by sending a telegram to the Yale Peabody Museum, which led to woefully inadequate records. Because of the considerable confusion regarding the Colorado localities, the author is reluctant to ascribe a species discovery to either Dinosaur Ridge or Garden Park. Suffice it to say new dinosaur discoveries by Marsh and Cope at Dinosaur Ridge and Garden Park include Apatosaurus, Camarasaurus, Diplodicus, Allosaurus and Stegosaurus.

DINOSAURS OF THE MORRISON AND GARDEN PARK AREAS:

Show pictures of these:

Diplodocus: Slender build. Long neck and whip-like tail. Length: 88 ft. Weight: 10-11 tons.

Apatosaurus ("Brontosaurus"): Stocky, more sturdy in build and not as long as Diplodocus. Weight: 30 tons.

Stegosaurus: Plates on back, horns on tail. Length: 20-24 ft. Weight: 1.5 tons or more.

Allosaurus: *Carnivore*, slightly smaller than the Cretaceous Tyrannosaurus Rex. Length: 39 ft. Weight: 3-4 tons.

In July, 1877, huge numbers of dinosaur bones were discovered at Como Bluff, Wyoming. Subsequently, the focus of dinosaur fossil exploration shifted from Colorado to Wyoming.

WESTERN COLORADO:

GRAND JUNCTION AREA:

In 1900, Elmer S. Riggs, Assistant Curator of Paleontology from what is now the Chicago Field Museum of Natural History, began collecting dinosaur bones from a

Box 6. Continued

hill (Riggs Hill), located three miles west of Grand Junction.   Riggs discovered a complete skeleton of Brachiosaurus, the largest known dinosaur at the time. Unfortunately, Riggs Hill was never protected and amateur collectors stripped it of bones.   In 1901, Riggs also excavated at Dinosaur Hill located eleven miles west of Grand Junction.   Today, Dinosaur Hill is partly owned by the Museum of Western Colorado and is open to visitors.

The Fruita Paleontological Research Area located three miles southwest of Fruita is a currently active site of excavation.   It has yielded fossils of reptiles. mammals and many dinosaurs.   One of which is an adult dinosaur no larger than a chicken!

Still further to the west of Grand Junction lies the 280 acre Rabbit Valley Research Area.   Along the 1.5 mile "Trail Through Time" quarries with exposed dinosaur bones can be viewed.

In Grand Junction, Dinosaur Valley Museum contains excellent displays of dinosaurs, fossilized dinosaur eggs and casts of dinosaur tracks.

At Dry Mesa Quarry south of Grand Junction and twenty miles west of Delta, bones of an extremely large Jurassic dinosaur named "Supersaurus" were found by scientists from Brigham Young University.   Studies of its bones indicate it was 54 feet tall, over ninety feet long and weighed eighty tons!   Later, another giant *herbivore*, Ultrasaurus, were found.   It appears to have been over 98 feet long with a weight of 130 tons.   Both have yet to be scientifically described and formally named.   Preparation of dinosaur bones for scientific study can require many years.   The Dry Mesa site was discovered in 1971 by Daniel E. Jones and his wife, Vivian.   James Jenson, a curator in the Geology Department of Brigham Young University (BYU), directed excavations at Dry Mesa for many years and was followed by Kenneth L. Stadtman, also of BYU (See Miller, et al, 1991). Dry Mesa Quarry may be visited, but not at all times.   The site is closed to the public except when excavation is in progress.

It is difficult to comprehend the existence of such enormous animals on earth but the bones do not lie!   The food requirements for large dinosaurs must have been enormous.   Vegetation on the Morrison landscape must have been abundant.

Box 6. Continued

DINOSAUR NATIONAL MONUMENT:

Dinosaur National Monument is located in northeastern Utah and northwestern Colorado, where in 1909, the Carnegie Museum discovered dinosaur bones in the Morrison Formation. Today the site is a national monument. Many excellent dinosaur bone specimens have been excavated. The best exposure, which visitors may view, contains a large number of bones still embedded in the sandstone where they were found. The site is housed in a permanent structure built to protect it. Well worth a visit. The Dinosaur National Monument is discussed in greater detail in the Parks and National Monuments area of this book.

NORTHERN NEW MEXICO:

In the Morrison of Northern New Mexico, the bones of a dinosaur tentatively named Seismosaurus have been found. From the size of the bones, Seismosaurus is judged to have been 150 feet long with a weight of approximately 100 tons! The average African elephant ranges from two to four tons and eats the equivalent of about two trees per day. How many trees would this immense dinosaur have eaten per day?

GASTROLITHS:

In addition to the occurrence of dinosaur bones in the Morrison Formation, occasionally "*gastroliths*" have been found (See Figure 31). These are "gizzard stones." As birds do today, some dinosaurs, probably *herbivores*, ingested stones which ground up their food. Gastroliths are highly polished pebbles, commonly composed of *chert* and ranged in size from "bird's egg" size to "hen's egg" size. The chert pebbles are exotic. No local geologic source existed on the Morrison landscape in Colorado to provide them.

Since there was no local source for these stones, it has been suggested the dinosaurs had to travel to other areas where they could find them. The nearest source of chert during Jurassic time was in western Utah where outcrops of Paleozoic limestones containing chert *nodules* were eroding. This is evidence that dinosaurs may have migrated, possibly seasonally.

Figure 31. Gastroliths. These "gizzard stones" from the Morrison Formation are composed of chert. The largest is 2.5 inches long.

WESTERN COLORADO:

Curtis Formation: Greenish gray *marine mudstones* with fossils.

Summerville Formation: *Nonmarine* red *shale* and *siltstone* with *gypsum*. Gypsum is a *marine evaporite* and is generally believed to have formed in shallow *lagoons* which oceanic waters occasionally entered.

Morrison Formation: Similar to that of eastern Colorado.

CRETACEOUS PERIOD: 138-65 *M.Y.A.*

The interior of the North American Continent began to lower. Oceans began to move inland from the Arctic and from the Gulf of Mexico area. These seaways met in southeastern Colorado and southwestern Kansas. The sea then began to slowly deepen and spread east and west. Eastern Colorado was inundated first and as time passed, the shoreline moved west over Colorado. Along the shoreline, *beach*, *barrier island* and *deltaic* sediments were deposited (See Figure 32).

LOWER CRETACEOUS:

EASTERN COLORADO:

Lytle Formation: The land tilted to the east due to moderate mountain-building in Utah. More and stronger rivers than those of the previous Morrison landscape appeared. Extensive river channel *sandstones* were deposited. The lower contact of the Lytle with the underlying Morrison is, by convention, defined as the base of the lowermost sandstone which contains small *chert* pebbles. Sandstones of the underlying Morrison do not contain chert pebbles. The source of the chert pebbles was from Lower Paleozoic *limestones* and *dolomites* exposed in the uplifted areas of western Utah. Streams draining eastward carried pebbles into Colorado. Chert occurs as *nodules* and beds in Lower Paleozoic carbonates. In Colorado, during deposition of both the Morrison and Lytle Formations, the landscape had low relief. There were no outcrops of Lower Paleozoic carbonates and therefore no local sources for chert. In earlier Morrison time, the landscape was nearly flat and streams could not move pebbles. As a result, Morrison sandstones do not contain chert.

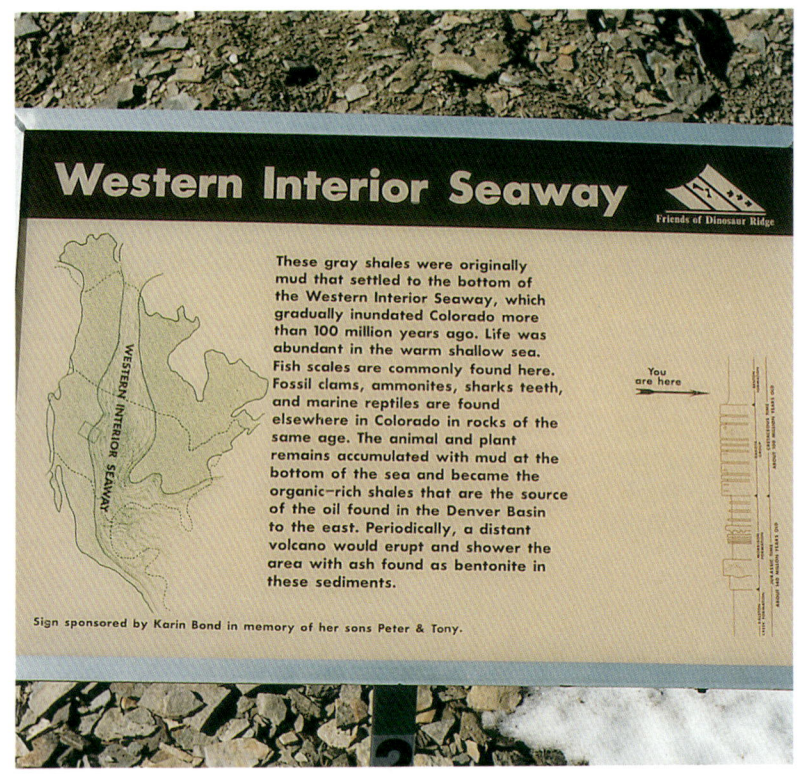

Figure 32. The Cretaceous Seaway. This is one of a series of informative signs placed on Dinosaur Ridge east of Red Rocks Park. It portrays the "Western Interior Seaway", another name for the Cretaceous Sea which invaded the interior of the North American continent during Upper Cretaceous time. Beginning about 110 million years ago, the interior of the continent began to lower. Seas moved inland from the vicinity of the Arctic Ocean and the Gulf of Mexico. These two arms of the oceans joined and then spread out. They formed a shallow ocean through the center of the continent. Colorado was ultimately covered. For a period of about thirty million years, the North American Continent was divided into two smaller continents.

78

South Platte Formation: As the shoreline moved west over Colorado, sediments were deposited in *beaches*, *barrier islands*, bays and in *deltas*. The South Platte, along with the Lytle and Morrison, are exposed in the Interstate 70 roadcut west of Denver where the highway passes through the Dakota *Hogback* (See Figure 33). Dinosaur tracks are abundant in the South Platte sandstones at Dinosaur Ridge (See Figure 34). The South Platte is composed of *shales* and *sandstones*.

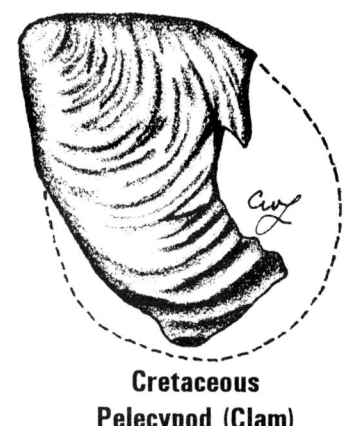

Cretaceous Pelecypod (Clam)

WESTERN COLORADO:

Dakota Sandstone: Rocks equivalent in age to the Lytle and the South Platte are present in western Colorado. These rocks were deposited west of the advancing shoreline and are largely *nonmarine* channel *sandstones*.

TRANSGRESSION OF THE CRETACEOUS SEA

As time passed, the shoreline continued to advance westward toward Utah. Ultimately the Cretaceous ocean covered all of Colorado and the eastern two-thirds of Utah (See Figure 32). This shallow ocean remained over Colorado for approximately 30 million years. Great thicknesses of *marine shale* along with lesser amounts of *limestone* and *sandstones* were deposited.

Shale is formed by deposition of clay particles. These are commonly brought from the land by rivers which empty into the sea. The shales of the Cretaceous Sea were deposited over millions of years, and during these times, the ocean would have contained abundant clay and would have been muddy or "dirty". At times, clay was not introduced into the sea and it "cleaned up". When ocean water is clean, organisms such as clams can thrive and create limestones. Marine sandstones were deposited along the shoreline as the ocean advanced and later retreated.

UPPER CRETACEOUS:

EASTERN COLORADO:

Mowry Shale: Thin, flaggy *siliceous shales*. Fossil fish scales and bones are abundant. Ammonite fossils also occur. The Mowry marks the beginning of the Upper Cretaceous.

Benton Formation: Dark gray *marine shales* with fossils, including ammonites, fish scales and bones, and shark's teeth. Deposited in an ocean some tens of feet deep.

Niobrara Limestone: White, chalky *marine limestone* and *calcareous shale*. Contains abundant, large *clam* fossils and sparse shark's teeth (See Figure 35).

Cretaceous Ammonite

Pierre Shale: Gray *marine shale*. Many thousands of feet thick. *Calcareous concretions* are abundant in the Pierre and they frequently contain *ammonite* fossils.

WESTERN COLORADO:

Mancos Shale: *Marine shale*. Many thousands of feet thick. Contains *ammonite* fossils. Forms lower slopes of the Book Cliffs and the south side of Grand Mesa.

Cretaceous Ammonite

80

Figure 33. Interstate 70 Roadcut. Jurassic and Cretaceous rocks are well exposed in the Interstate 70 roadcut west of Denver. As a result of uplifting of the mountains to the west, the layers of rock were tilted to the east. In the photo, the Cretaceous South Platte Formation is on the right side and the Cretaceous Lytle Formation on the left. Still farther to the left, Jurassic rocks (not shown) are exposed in the roadcut (the Morrison Formation and some of the Ralston Creek Formation). Although a spectacular display of rocks, the contacts between the various formations are difficult to determine at this locality.

81

Figure 34. Dinosaur Foot Tracks. Dinosaur foot prints
visible at the famous track site located on Dinosaur
Ridge near Morrison.   These tracks are preserved on
sandstone beds in the Cretaceous South Platte Formation.
They have been shaded by charcoal for better visibility.
The larger set of tracks is believed to have been made by
an adult Iguanodon. The smaller footprint in the lower
left was made by a juvenile Iguanodon.

REGRESSION OF THE CRETACEOUS SEA:

The middle of the continent began to rise in the late Cretaceous and the sea became shallower. The western shoreline began to retreat from Utah and slowly progressed eastward across Colorado. *Marine shales* continued to be deposited east of the shoreline. *Nonmarine* coastal plain sediments were deposited west of shoreline.

EASTERN COLORADO:

Fox Hills Sandstone: *Sandstones* and *shales*. The lower part of the Fox Hills is composed of several layers of sandstone which weather to a light yellow color. The sandstones are not *cross-bedded* and are parallel-laminated. Several thick layers of sandstone are finer-grained at the base and are slightly coarser-grained at the top. Such reverse grading of grain sizes in sandstones is characteristic of *beaches* or bars. These layers of sandstone have gradational bases and grade downward into the Pierre Shale. They are overlain by shales and are probably off-shore bars which formed in shallow ocean water seaward of the shoreline.

Laramie Formation: *Sandstones*, *shales* and coals. Similar in some respects to the Mesaverde Group of Western Colorado. Major differences include fewer *swamps* and less coal deposits and no *barrier islands*. The lower member of the Laramie is yellow to white sandstone, which in some areas contains <u>Ophiomorpha</u> *burrows*. These were created by a paleoshrimp (See Soister, p.223) and are known to occur dominantly along paleoshorelines and most particularly on sand beaches. Therefore, the lower Laramie sandstone is interpreted as a sand *beach* deposit. Fossils of palm tree fronds are occasionally found in these sandstones. Palm trees do not grow in climates where freezing temperatures are common. This indicates the climate was, at the least, semi-tropical.

The upper member of the Laramie is composed of interbedded *sandstones*, *siltstones* and *shales* with some coal lenses. These rocks were deposited on *floodplains* landward of the retreating Cretaceous Sea. Coals were deposited in *swamps* existing in places on the floodplain.

The *nonmarine* Laramie Formation is Late Cretaceous in age, therefore evidence of dinosaurs should be present. In fact, *fossil* bones of Tyrannosaurus Rex have been found in Laramie rocks in the Denver metro area. Tyrannosaurus Rex was indeed present in eastern Colorado during the Late Cretaceous.

Figure 35. Niobrara Limestone. An aerial view of the Red Rocks Park/Morrison/Dakota Hogback area. Direction of view is to the north. The photo was taken well before present day development. The white ridge on the east (right) side of the Dakota Hogback is the Niobrara Limestone. After the picture was taken, the Niobrara was mined and turned into cement or lime soil conditioner. It is completely gone from the surface of this area today. Other features of interest include: Red Rocks Park, left center. The famous amphitheater had not been built at this time; the tiny village of Morrison visible below Red Rocks Park; the Dakota Hogback (Dinosaur Ridge is at the far end of the ridge) and South Table Mountain, upper right. Photo by T.S. Lovering, no date. USGS Photo Library.

WESTERN COLORADO:

Mesaverde Group: In some areas, the Mesaverde Group is subdivided into several formations. In other areas, it is not and is simply referred to as the Mesaverde Formation. It is composed of *sandstones*, *shales* and coals. The Mesaverde is extensively exposed in western Colorado (See Figure 36). These rocks were deposited on *nonmarine* coastal plain and *marginal marine depositional environments* as the Cretaceous Sea retreated eastward. *Barrier islands* which today contain *fossil clam* and shrimp *burrows* existed along the ancient coastline. Coal-forming *swamps* were common on the landward side of the islands. The climate was humid and semi-tropical with palm trees which have been preserved as fossils (See Figure 37). The Mesaverde in western Colorado contains very large reserves of coal.

Figure 36. Chimney Rock. This scenic butte is located southwest of the Mesa Verde National Park. It can be viewed from the highway south of Cortez. The upper cap is composed of Cretaceous Mesaverde Formation sandstones underlain by the easily erodible shale of the Mancos Formation.

85

Figure 37. Fossil Palm Tree Frond. This fossil palm tree frond was found in the Coryell coal mines at Newcastle, Colorado, west of Glenwood Springs. The fossil is on sandstone from coal-bearing rocks equivalent to the Mesaverde Group of Late Cretaceous age. It is evidence of a semi-tropical or possibly a tropical climate in Colorado at this time. Such palm tree fossils are not uncommon in Late Cretaceous and Early Tertiary rocks. Photo by H.S. Gale, no date. USGS Photo Library.

THE ROCKY MOUNTAINS ARRIVE

The modern-day Rockies began to rise near the end of the Cretaceous Period approximately 72 *M.Y.A.* Much of the uplifting involved vertical movements (*fault block mountains*) creating high mountain ranges and *intermontane* basins (valleys). The tectonic forces involved were provided by the collision of the westerly-moving North American Plate with the easterly-moving Pacific Plate. These mountain-building episodes are named the *Laramide Orogeny*. They continued on into the Late Tertiary creating the Rocky Mountains of today. Initially, much of the state was still near sea level with a semi-tropical climate. At this time, the North American continent was in the same approximate latitude as today, however, the earth was warmer. There were no polar ice caps. Dinosaur bones of Late Cretaceous age have been found in the Arctic, and other fossils prove palm trees grew near Anchorage, Alaska!

EASTERN COLORADO:

Arapahoe Formation: Composed of *sandstones* and *conglomerates* deposited in *alluvial fans*. The conglomerates contain rounded *chert* and *quartz* pebbles. *Limestone* pebbles and small cobbles and other *sedimentary rock* types are also present. All of which originated from the *erosion* of Mesozoic and Paleozoic rocks that topped the initial uplifts of the future Rocky Mountains. The pebbles and cobbles are the very first evidence of mountain-building processes in Colorado since the Pennsylvanian Period when the Ancestral uplifts were created.

Denver Formation: The Denver Formation is composed of alluvial sediments which built eastward, from the newly rising mountains, into the plains. *Weathering* and *erosion* plus extensive volcanic action left *claystone*, *siltstone*, *sandstone* and *conglomerate*. Many of these deposits are comprised of *andesitic* materials derived from erosion of *volcanic rocks* (created as the Rocky Mountains rose). Conglomerates of the Arapahoe and the Denver formations are the **first major** conglomerates deposited since the Ancestral Rocky Mountains. They provide important geological evidence for the initial uplift of our modern mountains. These sediments were *alluvial fan* and *bajada* deposits which formed on the flanks of the rising and eroding mountains (largely volcanic at this time).

WESTERN COLORADO:

Many *alluvial fan* deposits in *intermontane* basins. Some of these have local formational names but many of the deposits are unnamed.

CENOZOIC ERA:

The end of the Mesozoic and the beginning of the Cenozoic is marked by *extinction* of 75% of all species living on earth at that time, including dinosaurs. There have been many *mass extinctions* on earth through geologic time, and some of these were used to establish boundaries in the Geologic Time Scale. Until recently, the most common explanation for the cause of such events of mass death involved climatic change. In 1978, a new theory was set forth which proposed the 75% mass extinction at the end of the Mesozoic was caused by the effects of a massive meteorite impact. A thin layer of *clay* present at the Cretaceous-Tertiary time boundary (known amongst geologists as the "*K-T Boundary*") contains significant amounts of the rare element, *iridium*. Additionally, high-velocity *shocked quartz* and *elemental carbon* are contained in the clay. This evidence found in the thin clay layer (often referred to as the impact layer) is consistent with the known chemistry of meteorites and characteristics of their collisions. For a detailed discussion of the Meteorite Impact Theory and the significance of the impact layer and its composition, see Box 7.

BOX 7.   THE METEORITE IMPACT THEORY

Today, we are very concerned with the modern *extinctions* of animal and plant species on earth, and deservedly so. It is a paradox that approximately 99% of all species living on earth through geologic time have become extinct. Geologists know this from the study of fossils. Species evolve, flourish and then disappear, their fossils never to be seen again in younger rocks.

There are certain places in the *fossil* record where many organisms disappear at the same time. These are called *mass extinctions*. Geologists have known about mass extinctions for a very long time. In fact, mass extinctions played an important part in the creation of the Geologic Time Scale, which we still use today. The Geologic Time Scale was originally devised using relative age-dating methods. Many of the time boundaries were placed at points in the rocks where mass extinctions of species occurred.

There have been many mass extinctions in the record of life on earth. In terms of numbers affected, some were large, some were moderate in size and some were small. The largest occurred at the end of the Paleozoic Era when 95% of all species living on the land and in the sea

---

Box 7. Continued

became extinct. In fact, life on earth almost ended at that time. From the 5% that survived, the life forms of the Mesozoic evolved, including the magnificent dinosaurs. The next largest occurred at the end of the Mesozoic Era when 75% of all species died, including the dinosaurs.

How can such mass extinctions be explained? The fossil record of mass extinctions is global in scale and is evident on continents and in the oceans. Hence, whatever the explanation is, it has to include all of the earth at the same time. This is, and has been, an argumentative subject and there still is no general agreement among geologists. For many years, a fairly well-accepted theory was climatic change. Periodically, the world's climate was thought to have materially altered and this was believed to be responsible for the mass extinctions.

Recently, some startling and indeed amazing information has come to light which indicates mass extinctions may be due to much more spectacular events than mere climate changes! In 1979 at an outcrop in Italy, Walter Alvarez, a geologist, collected samples from a one and one/half inch *clay* layer which separates Cretaceous from Tertiary *marine limestones*. Together with his father, Luis Alvarez, a Nobel physicist, they conducted laboratory investigations which revealed that this clay layer is anomalously rich in the heavy element, *iridium*. Iridium is a very rare element in the earth's *crust*, so the earth itself cannot be a source for the iridium in the clay layer, but some meteorites are known to be rich in iridium. Based upon their findings, the two scientists proposed that the earth was struck by a large iridium-rich meteorite 65 million years ago. This would explain the extinction of 75% of all living things on earth.

Meteorites travel at high velocities, 50,000 mph or even faster! From studies of known large meteorite craters, scientists have learned large meteorites explode on impact. The heat and pressure of such high velocity collisions vaporizes the meteorite and blasts a large crater. The cataclysmic explosion puts enormous quantities of heat, dust, and rock vapor into the earth's atmosphere. The heat released creates a large fire ball which would roll out from the impact site.

Walter and Luis theorized that the dust and rock vapor in the atmosphere blocked sunlight for enough time to interrupt the food chain. This caused the death of most life forms existing at the time. The diameter of the

Box 7. Continued

meteorite was calculated to have been eight kilometers or larger, and such an impact would blast a crater 110 kilometers wide. The explosion would have been 1,000 times greater than the potential explosive strength of the entire United States and former Soviet Union nuclear arsenals (Muller, 1988, p. 64).

Other geologists were quick to criticize this theory. Obviously, if the earth were struck by a large iridium-rich meteorite, then iridium should be found at the Cretaceous-Tertiary boundary elsewhere in the world. But where was this evidence? Geologists immediately began collecting and analyzing samples from other places. As of 1986, iridium has been found in samples from the Cretaceous-Tertiary time boundary at eighty localities around the world.

Another group of scientists discovered the clay layer at the Cretaceous-Tertiary time boundary is anomalously rich in *elemental carbon*. This is what you and I know as "soot" and is derived from smoke. In fact, the isotopic nature of the elemental carbon is the same as that obtained from the burning of trees. In light of this knowledge, they theorized that the enormous fire ball caused by the meteorite "blast" set many of the earth's forests on fire. This put large quantities of smoke into the atmosphere, further blocking sunlight for a period of time.

Still other scientific analyses revealed microscopic bits of *quartz* in the layer of clay. These contain a very unusual *high-velocity shock fracture*. The shock fracture has a "herringbone" pattern. We know of only two ways it may be created. One is by atomic bomb blasts. Atomic bombs explode with a very high velocity and quartz containing the high-velocity shock fracture has been found in blast craters. The only other way high-velocity shock fractures may be created in quartz is by meteorite impacts. Quartz containing these fractures has been found in the craters of known meteorite impacts.

Subsequently, paleontologists (geologists who specialize in the study of fossils) have plotted known mass extinctions on a detailed linear time scale. They found mass extinctions occur in a cyclic pattern every twenty-six million years! Some of these extinctions are larger than others, but none-the-less they all are extinctions. It was then proposed every twenty-six million years, the earth is struck by one or more meteorites, causing extinctions. Again scientists were critical. They commented *iridium* should then be found in the geologic record of other mass

Box 7. Continued

extinctions. As of 1988, five mass extinctions are now known to have iridium signals (Muller, 1988, p. 184).

At this point, the question is undoubtedly being asked: "Where are we today in this cycle of mass extinctions?" Fortunately, it will be another twelve to thirteen million years before the earth is scheduled for its next mass extinction!

Upon review of this remarkable information regarding <u>cyclic</u> mass extinctions, other scientists entered the "fray". They pointed out that our Sun, which is a star, is an unusual star. Many of the stars in the universe come in pairs, known as "companion stars" or sometimes as "binary stars" and they orbit around each other. Our Sun (star) is unusual because it does not have a companion. Or does it? These scientists noted there is a class of stars known as *"dim dwarf stars"*. Although small, they are very dense with high gravity and emit only small amounts of light. They proposed that our Sun may have a "dim dwarf" companion with an orbital period of twenty-six million years. The scientists nicknamed the theoretical "dim dwarf" companion star, *"Nemesis, the Death Star"* (Nemesis is the ancient mythological Goddess of retributive justice or vengeance). This immediately caused trouble with peers as there are some very strict conventions for naming celestial bodies and levity is not one of them! Be that as it may, the nickname "Nemesis, the Death Star" appears to have stuck, at least informally.

It was further pointed out that there is an enormous cloud of *comets* located beyond Pluto. This cloud of comets is known as the *Oort Cloud*. It contains thousands of comets, many of which are quite large with rocky cores. They proposed when "Nemesis" comes by orbiting the Sun, its gravity would disrupt some of the comets in the Oort Cloud. These disrupted comets would then fall back toward the Sun and strike the planets and the Sun. The comet(s) striking the earth would cause mass extinctions. If "Nemesis" exists, then where is it? Astronomers have been reviewing astronomical records and calculating orbits in an attempt to find a suitable candidate. To date, none has been found. We may have to wait twelve or thirteen million years to find it! The Nemesis Theory may be, after all, just a theory.

Still another obvious question is where did the meteorite (or asteroid or comet) strike the earth at the end of the Cretaceous Period?" The thickness of the impact

Box 7.   Continued

layer containing the high-velocity *shocked quartz* increases dramatically as you approach an area in the Gulf of Mexico. A very large (110 mile diameter) buried meteorite impact crater has been identified on the tip of the Yucatan Peninsula of Mexico.   It is exactly the right age (65 million years old).   This is known as the Chixublub impact crater (Hildebrand, 1993).   Many scientists now believe this is the place (or at least one of the places) where Earth was struck by a very large meteorite at the end of the Cretaceous.

The Yucatan Peninsula was under the ocean 65 million years ago.   The impact of a five-mile diameter meteorite would have created a giant sea wave which would have rapidly moved out from the meteor crater.   This wave could be expected to rip up the sea floor where it impacted. Indeed, ripped-up *limestones* immediately below the *K-T boundary* are found in coastal areas of Texas and in other areas around the Gulf of Mexico.

As discussed above, when giant meteorites impact a large immovable object, they explode releasing an enormous quantity of energy.   This energy, which is heat, creates a large fireball which rolls out at extreme velocities from the impact site.   During the summer of 1994, the impact of the Shoemaker-Levey Comet #9 fragments on Jupiter created giant fireballs, at least one of which was larger than earth!   The Chixublub impact is estimated to have created a 3,000 kilometer fireball which would have rolled across what we now call the United States into southern Canada. In other words, Colorado burned!   Everything combustible such as vegetation and any organic matter (dinosaurs?) would have been consumed.   If this were to happen today, in a fraction of a second, every thing combustible (including humans) would flash to smoke!

Other recent scientific studies of *microtektites* (formed by the splash of molten rock from meteorite impacts), *smectite clay* and *osmium* isotope ratios in the Cretaceous-Tertiary boundary layer, support the meteorite impact theory.   Furthermore, it appears that a piece of the Chixublub meteorite has been found.   University of California geochemist Frank Kyte examined a core drilled in the northwest Pacific Ocean.   Kyte found a chip of meteorite in a 65-million year old layer of *iridium*-rich mud in the drill core.   The meteorite fragment was rich in metallic elements, including iridium.   Kyte believes it is a fragment of the Chixublub meteorite (Kyte, 1996).

Box 7. Continued

Is there any evidence Colorado burned 65 million years ago? The answer is yes. At the time, most of Colorado consisted of uplifted areas (the early, rising Rocky Mountains). They were eroding and creating vast *alluvial fans* which covered much of the state, except for an area near Trinidad in southern Colorado known as the Raton Basin. Alluvial fans are highly erosive sites and the Cretaceous-Tertiary boundary layer was apparently eroded away and lost. The Raton Basin, which also includes part of northern New Mexico, was an area of *swamps* and marshes which are sites of deposition with no *erosion*. The Cretaceous-Tertiary boundary material fell into the swamps and was preserved as two centimeters of shale (the impact layer) (See Figure 38). This layer has been found in many sites in the basin and it contains *iridium*, *shocked quartz*, and *elemental carbon*. Studies of *fossil pollen* and spores from this layer show immediately before the impact, many different types of plants grew in the area. Immediately above the impact layer, there is a great reduction in pollen from these plants, indicating that many of them ceased to exist (burned?). Instead, there is a high concentration of fern spores (Nichols, et al, 1990). Ferns are known as the great recolonizers after a fire! The study of the fossil pollen and spores indicates Colorado burned at the time of the impact of the Chixublub meteorite.

There is still another most serious question for the reader to ponder regarding large meteorite impacts on earth. Until recently, none of the information obtained from the studies of the impact at Chixublub and the presumed mass extinction which followed gave any indication of the time frame involved. How long afterward was life on earth adversely affected? Was it a period of months or years or even longer periods of time?

In February of 1997, scientists on the drill ship Joides Resolution collected sediment cores on the ocean floor off the east coast of Florida. From the cores, they found abundant *fossil* evidence of life in the sediments <u>below</u> the Cretaceous-Tertiary Boundary. In other words, before the impact of the meteorite at Chixublub, the ocean teemed with life. At the Cretaceous-Tertiary Boundary, they found a layer containing small green glass pebbles (*microtektites*) and above, a rusty brown layer (the impact layer). On top of this was a sparingly *fossiliferous* two-inch layer of gray clay indicating a nearly dead sea. Above the barren two-inch layer of clay, samples show evidence of renewed life. The ocean where the cores were

Box 7. Continued

taken is 8,500 feet deep. At these depths, the rate of deposition of clay is very slow, approximately 10 millimeters/thousand years. At this slow rate of deposition, it might have taken 5,000 years or so for the barren clay to be deposited. This is not good news. It implies after the Chixublub impact, life in the ocean was seriously impacted for 5,000 years. Therefore, if the earth were struck by a large meteorite today, the very survival of humans might be in doubt! 5,000 years is a very long time to wait for life on earth to return to some semblance of normalcy (Recer, 1997).

At this point, the difference between facts and interpretations or theories should be clarified. Facts are things like iridium, elemental carbon, high-velocity shocked quartz and the twenty-six million year extinction cycle. Interpretations and/or theories are how scientists explain the meaning of facts.

Most geologists now agree there is convincing evidence for the impact of an object from outer space at the end of the Cretaceous. Still, not all of them believe this impact caused the world-wide mass extinction which occurred at the same time.

What we have is the classic court-room drama of dead bodies and a smoking gun! These are facts. It is still an interpretation or theory to say that the smoking gun <u>did kill</u> the dead bodies. Scientists still do not have <u>proof</u> the mass extinctions are caused by meteorite impacts, but it is not an unreasonable interpretation of the known facts. Research concerning this remarkable theory is still continuing. Undoubtedly more startling revelations will be made in the future.

The K-T Boundary is difficult to locate in much of Colorado as most of the state was covered by *alluvial fans* at this time. In a highly erosive, *nonmarine* alluvial fan system, the original rock layers containing the impact layer would have been eroded away. In other words, a small *unconformity* exists in the rocks in many areas of the state.

In eastern Colorado, the K-T Boundary is somewhere near the middle of the Denver Formation. Apparently, as elsewhere in the state, an unconformity is present. In the Raton Basin (a sedimentary basin) in southern Colorado and northern New Mexico, *swamps* and marshes existed, which are sites of depositional *accretion*. In other words, whatever falls into them will be preserved. Consequently, the exact layer deposited at the end of the Cretaceous is present in the geologic record of the Raton Basin (See Figure 38). Samples of the impact layer from the Raton Basin contain *iridium*, *shocked quartz* and *elemental carbon*. *Paleopalynology* studies from samples taken just below the impact layer reveal *fossil pollen* which indicate that many plants and trees grew in the swamps. Rock samples from immediately above the impact layer indicate that a number of plant species had disappeared as evidenced by the absence of pollen. Fossil fern spores are in great abundance above the impact layer (See Nichols, et al, 1990). Ferns are known as the great recolonizers after fires. Geologists interpret this as evidence of a giant fireball from the impact which rolled across Colorado. **Colorado burned at this time!** Should this happen today, everyone and everything in the state would be destroyed.

TERTIARY PERIOD: 65-2 *M.Y.A.*

Tertiary rocks over much of Colorado are largely *nonmarine alluvial* deposits, with many *volcanic ashes* and *lava flows*. There are numerous localized formational names. In many areas, rocks are simply called "Tertiary". In the northwest portion of the state, significant "*oil shales*" were deposited in a large shallow lake.

EASTERN COLORADO:

Denver Formation: In the early Tertiary, the Denver Formation continued to be deposited until the end of the Early Paleocene Epoch.

Green Mountain Conglomerate: Outcrops only on upper slopes of Green Mountain near Morrison. This formation is of Late Paleocene age. The Green Mountain is primarily comprised of *claystone*, *siltstone* and *sandstone* with layers of *conglomerate* consisting of pebbles, cobbles and boulders of *gneiss* and *granitic pegmatite* along with lesser amounts of *basalt* and *quartzite*. This is in contrast to the *andesitic* nature of the underlying Denver Formation. Large petrified tree logs, some of which may be redwoods, are occasionally found in the Green Mountain Conglomerate. These rocks were deposited as *alluvial fans* which may have coalesced to form a *bajada* or *piedmont*. At the time of the deposition of the Green Mountain Conglomerate, the older *sedimentary rocks* and the early volcanics had been eroded off the top of the rising Rocky Mountains. Precambrian-age granite and gneiss are now exposed on the higher summits.

Dawson Arkose: The lower Dawson is of Late Paleocene age and the upper part is Eocene. Although separated geographically, the lower Dawson and the Green Mountain Conglomerate are apparently *stratigraphically equivalent units*. The lower Dawson is composed of *claystone*, *siltstone* and *conglomerate*. The upper Dawson is mostly conglomerate with lesser amounts of claystone.

Figure 38. Cretaceous-Tertiary Impact Layer. The thin white layer exposed horizontally in the photo is the impact layer created when a meteorite estimated to be five miles in diameter impacted in Mexico. This layer is preserved at the K-T Boundary in rocks of southern Colorado and northern New Mexico. The site of the picture is in a roadcut on Interstate 25 between Trinidad and Raton. The impact layer in this area was discovered and studied by Charles Pillmore and associates of the United States Geological Survey. The layer contains iridium and high velocity shocked quartz. A rock hammer, visible on the left, provides scale.

WESTERN COLORADO:

Numerous local formational names and many unnamed Tertiary deposits.

LATE EARLY TERTIARY (EOCENE):

EASTERN COLORADO: No rocks preserved.

WESTERN COLORADO:

Wasatch Formation: Comprised of basinal deposits of tan to pinkish-tan basinal deposits of *conglomerate*, *sandstone* and *shale*. The environment of deposition included rivers, *floodplains* and *alluvial fans*.

Green River Formation: During the Eocene, a very large fresh-water lake known as Lake Goshiute formed in this region, and *sandstones*, *shales* and shaley *limestones* were deposited in it. This formation contains very substantial amounts of "*oil shale*" which is rich in an organic matter known as *kerogen*. Oil may be obtained from these rocks by heating and distillation (See Figure 39). "Oil shales" look like ordinary grey-brown shales but they are composed principally of *calcite* plus *clay* and kerogen. So they are actually an organic-rich shaley limestone. These "oil shales" contain the largest reserve of petroleum in the United States. At the present time, mining and distillation of the rocks to obtain the oil is not economic. Excellent fish fossils occur in some of the lake deposits. When the Wasatch and Green River Formations were deposited, this area was a low basin. Today, it is known by geologists as the Piceance Basin. The current landscape, however, consists of uplifted *plateaus*.

LATE TERTIARY:

EASTERN COLORADO:

Ogallala Formation: *Conglomerates* and *sandstones* deposited by streams draining eastward from the mountains. The deposits were *alluvial fans* which coalesced to form a *piedmont* east of the mountains.

Castle Rock Conglomerate: *Conglomerates* and some *volcanic ash* deposits.

WESTERN COLORADO: *Volcanics* and various basin-fill *clastics*.

Figure 39. Green River Formation. This formation is well displayed in the Roan Cliffs west of the town of Rifle on Interstate 70. The upper part of the cliffs is composed of the Green River, which contains oil shales. The Wasatch Formation is exposed in the lower part.

MIDDLE TERTIARY VOLCANISM:

From 40 to 28 million years ago (Oligocene-Miocene epochs), extensive *volcanic* activity occurred in the San Juan Mountain area in southwestern Colorado. This was a huge volcanic area 100 miles across and contained several large *calderas*. Enormous amounts of volcanics were ejected in numerous very great volcanic explosions. At this time, most of Colorado would have been blanketed by ash deposits. Much of this ash was carried from the *volcanoes* by hot "*ash flows*" and farther away from the volcanoes, cooler ash fell from the skies. After the ash was deposited, *lahars* (mud flows) would have plagued the state for many years. Lahars are mud flows formed when precipitation or melt water mixes with the loose, sometimes still hot, dry ash. Remnants of these lahars are still present in many areas of the state. During the Tertiary, volcanism also occurred in other portions of the state (See Figure 40). Clearly, during the Tertiary volcanic episodes, Colorado was a very unpleasant place for major periods of time.

Figure 40.   Tertiary Volcanic Neck.   One of the most unusual and interesting volcanic necks in Colorado is located south of the town of Yampa. It is identified by a highway sign as "Finger Rock". In the photo, the neck is the vertical spire of rock and the white rock to the right of the neck is volcanic tuff. North of Yampa, several other volcanic necks may be seen, but none as dramatic as "Finger Rock." These volcanic necks are quite young, ranging from 7 to 10 million years in age.

Florissant Fossil Beds: During very late Eocene time, a mountain lake was formed in the vicinity of Florissant. It was created by *lava flows* which dammed an ancient stream. *Volcanic ash* periodically fell into the lake. A large portion of these lake beds today are located in the Florissant Fossil Beds National Monument. Large fossilized Redwood tree stumps in growth position and *fossil* tree leaves and insects have been found in the ash-rich lake *shales*. Some fossil fish and one marsupial also have been found. Elsewhere in the world, insect fossils are extremely rare. Their occurrence here is of great importance. The fossil tree stumps and fossil leaves demonstrate that large redwood groves existed here. Also present were palm trees, birch, willow. maple, beech, hickory, aspen and fir trees. Such an assemblage of trees indicates that the climate at this time was transitional between semi-tropical and temperate. At this time, Colorado had high mountains, but the land at the base of the mountains was still near sealevel. Huge redwoods were not confined to the Florissant area. As evidenced by abundant petrified wood specimens found in many areas, redwoods grew over much of Colorado and indeed the Rocky Mountain region. With the groves of large redwoods growing everywhere, Colorado may have been more beautiful than any other time in its history. Today, redwoods are almost extinct, with a few groves left in California. For more information, see Florissant Fossil Beds National Monument on page 158.

MINERALIZATION:

During Early and Middle Tertiary time, hot water (hydrothermal) solutions rich in gold, silver, zinc, lead, copper and iron seeped through and mineralized rocks. These fluids are believed to have been associated with *igneous* activity. Thus were created much of Colorado's rich *mineral* deposits. These mineral deposits largely were emplaced in a trend extending from Silverton northeastward to Boulder. This trend is known as the *Colorado Mineral Belt* (See Figure 41 and the MINING IN COLORADO section).

REGIONAL UPLIFT OF THE ROCKY MOUNTAIN REGION

Beginning approximately ten to fifteen million years ago, Colorado and the Rocky Mountain region were slowly raised by a general uplift (3,000 to 5,000 feet) to their present elevations. The cause of this uplift is unknown but it was probably related to plate tectonics. As the climate slowly cooled and became more arid, redwoods retreated to California. Colorado assumed its present character of beautiful, high alpine mountains, *plateaus* and high plains.

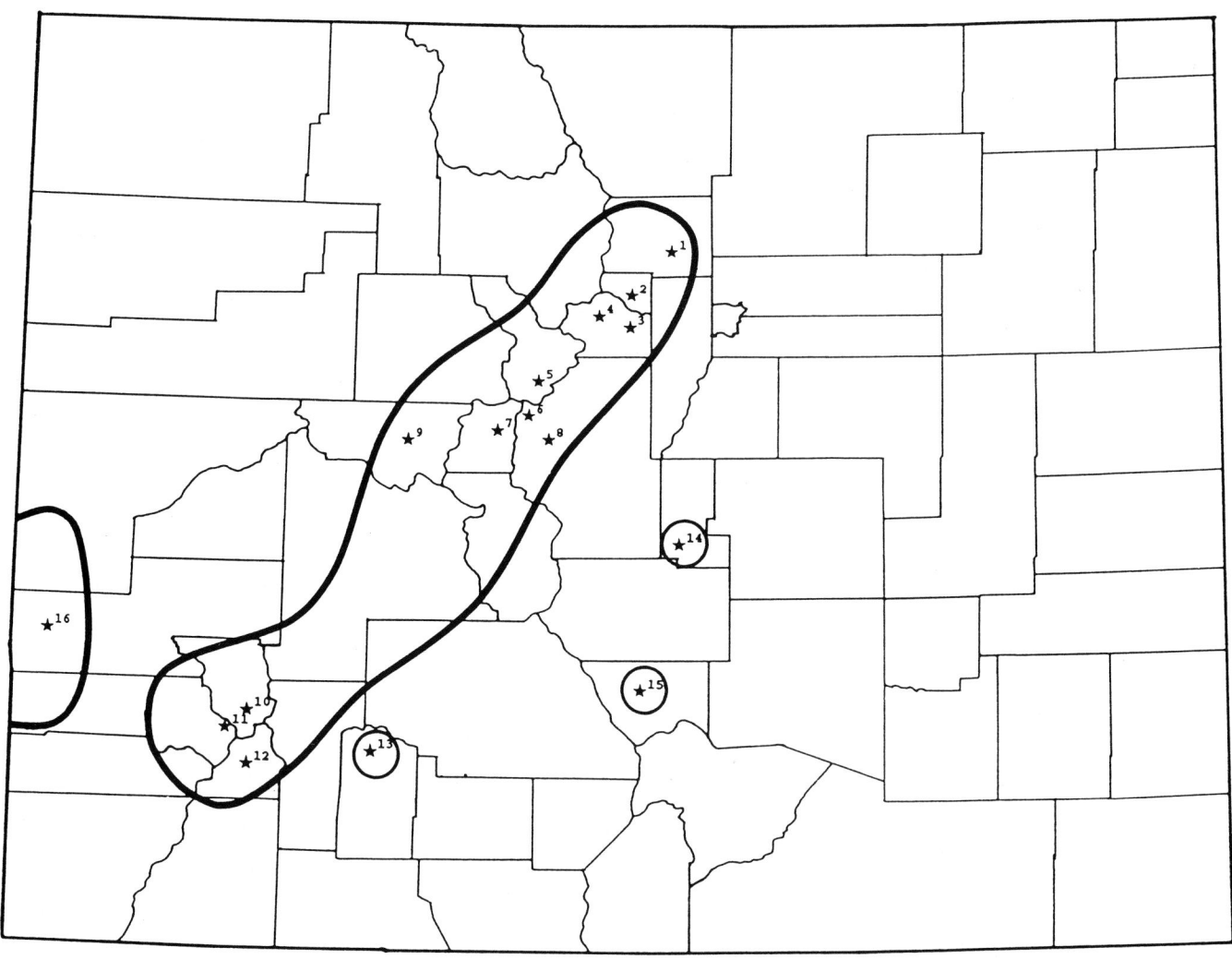

★[1]	Boulder		★[9]	Aspen
★[2]	Blackhawk-Central City		★[10]	Ouray
★[3]	Idaho Springs		★[11]	Telluride
★[4]	Georgetown-Empire-Silver Plume		★[12]	Silverton
★[5]	Breckenridge		★[13]	Creede
★[6]	Climax		★[14]	Cripple Creek
★[7]	Leadville		★[15]	Silver Cliff-Westcliffe
★[8]	Fairplay		★[16]	Uravan Mineral District

Figure 41. Colorado Mineral Belt. A majority of Colorado's major mineral deposits occur in a southwest-northeast trend across the mountains. Other mining districts not in the mineral belt are also shown.

QUATERNARY PERIOD: 2 *M.Y.A.* to present.

PLEISTOCENE EPOCH: 2 M.Y.A. to 10,000 yrs ago.

Approximately 2 million years ago, the earth cooled dramatically. Great *continental glaciers* covered most of Canada and much of the northern United States. Ice sheets did not extend into Colorado, but the climate was cooler. *Alpine glaciers* developed in many mountain ranges.

Glacial features: *Valley glaciers* extended down to 8,000 feet. *U-shaped valleys* were carved by the glaciers and many *cirques* (See Figure 42) are visible today at the heads of glacial valleys. Glacial deposits include lateral, medial and terminal *moraines*. Several small glaciers still exist in sheltered cirques above 11,000 feet. Among these are St. Mary's Glacier west of Denver, Arapahoe Glacier west of Boulder and several small glaciers in Rocky Mountain National Park. In Colorado, *permafrost* exists in *talus* deposits on the northern sides of mountains above 12,000 feet.

Mammoths and Mastodons: As evidenced by *fossil* discoveries, these large creatures were present in Colorado during the Cenozoic. They are believed to be related to modern-day elephants. The *mammoths* lived during the middle Tertiary. The *mastodons* were abundant during the Pleistocene. They had large bodies, upcurved tusks and considerable body hair. Many of the mastodon fossil sites also contain evidence of early humans, and it appears that the animals were either scavenged or killed by them. Some scientists speculate these large animals may have been driven to *extinction* by early hunters who killed and ate them!

Early Humans: Early nomadic hunters arrived in Colorado in Late Pleistocene time, possibly 20,000 years ago.

Holocene Epoch: 10,000 years to present.

The end of continental *glaciation* occurred about 10,000 years ago and the earth subsequently warmed. What we consider to be "normal" temperature is in fact abnormal. The earth's long-term average temperature throughout geologic age is somewhat warmer than what we have today (warm enough that there were no polar icecaps!). The earth has yet to return to its normal warmer ice-free conditions as large polar icecaps still exist. Some scientists believe the earth may return to a glacial episode in the near future (a chilling thought!). New evidence from ice cores from Greenland and from deep sea sediment samples indicates a return to a glacial stage could happen within a span as short as five years!

Figure 42. Cirques. A stunning aerial view of the glacial cirques above Berthoud Pass on U.S. Highway 40 south of the ski area of Winter Park. Note the unpaved road to the pass below. The summit of Berthoud Pass is the small treeless area with a wide spot in the road. It is in the right center portion of the picture. Photo by T.S. Lovering, no date. USGS Photo Library.

The last *volcanic* activity within Colorado occurred about 4,700 years ago near Dotsero, along Interstate 70 west of Eagle. A small *cinder cone* is visible north of the highway (See Figure 43) and a small *basalt lava flow* is visible south of the freeway. The *volcano* and lava flow are best viewed by exiting I-70 at Dotsero onto the service road.

Presently, the geologic processes active in Colorado are *weathering*, *erosion*, transportation, deposition (of sediment) and occasionally, a few relatively small *earthquakes*. Many areas of mountains are eroding by *landslides*. In geologic time, the primary mechanism whereby areas of high relief are reduced to low relief is by mass wasting, of which the primary mechanism is landsliding. Paleolandslide deposits have been mapped in many places in the mountains and *plateaus* of Colorado. In the state, the larger landslide deposits are shown on the Geologic Map of Colorado (in pocket). The map by Tweto (1976) shows many more.

Fortunately for most people (except possibly geologists!), the geologic processes operating in Colorado today are quite boring. Nothing exciting is taking place, such as volcanoes erupting, fireballs from meteorite impacts rolling past, glaciers grinding out of the mountains or major earthquakes from mountain building!

SUMMARY OF GEOLOGIC HISTORY

The rocks of Colorado tell a fantastic story of dramatic change throughout geologic time. Repeatedly, the state has been covered by oceans. Coastal shorelines have moved back and forth across the state. Mountain ranges have risen and fallen. Colorado has endured drastic climatic changes from tropical to temperate to glacial. The state has exhibited landscapes ranging from indescribably ugly to extremely beautiful. The reader should understand that landscapes are evolutionary and change through geologic time. Today, Colorado has a very beautiful alpine environment. Enjoy it while it lasts!

Figure 43. Dotsero Volcano. At 4,700 years old, this is the youngest volcano in Colorado. Age dating is based upon a Carbon 14 study of wood found under *scoria* in the associated lava flow. At the center of the photo, a small brown basaltic cinder cone is visible. The volcano is located near Dotsero north of Interstate 70 east of Rifle. Basalt lava flowed from the volcano and may be seen south of the freeway across from the volcano. Basaltic magma comes from the mantle of the earth. It is possible the magma at Dotsero may be related to continental rifting occurring in the Rio Grande Rift. The Rio Grande Rift extends up through the Arkansas River Valley. For more information on rifting and the Arkansas River Valley, see pages 130 and 132. Today, the cinder cone is mined for aggregate and may ultimately disappear. A modern form of erosion?

    A close-up view of some of the dinosaur tracks at
Dinosaur Ridge near Morrison. The larger ones are
thought to have been made by Iguanodon, a herbivore. The
small one with narrow toes is a theropod which was
approximately the size of an ostrich.

THIS LAND WE CALL COLORADO

Colorado is a political subdivision. To a geologist, Colorado is a landscape with geological features which know no boundaries. Colorado may be conveniently subdivided into three geographic and geologic provinces. These are the mountains, the *plateaus* to the west and the plains to the east. For reference to the geographic location of geologic features discussed in this section, see Figure 5, Location Map (in pocket).

MOUNTAINS

The majority of the present day mountains of Colorado were formed by vertical uplifts of *basement rocks* and are bounded by large, high-angle *faults* of great displacement. Some mountains were built by *volcanoes* and a very few by *overthrusting* (by compression of the earth's crust, older rocks are "thrust' over younger rocks along a low angle fault). Still others were created by the processes of *weathering* and *erosion*, that removed softer surrounding rocks, leaving a mountain composed of more resistant rocks. The mountain-building processes that created the Rocky Mountains of today began in Colorado approximately 72 million years ago. These geologic processes were caused by collision between the Pacific plate and the North American plate. This took place under what is now California and Oregon (they were under the ocean at that time). Compression from plate collisions caused the vertical uplifts in Colorado. Colorado's volcanic activity is believed to have come from melting of rock materials subducted under the continent (see Box 1. Plate Tectonics).

FRONT RANGE:

The Front Range extends from Canon City northward into Wyoming. These mountains are composed of Precambrian *granites* and *metamorphic rocks* which are largely *gneiss* (See Figure 4, in pocket). They are bounded by vertical *faults* on the east and west sides. West of Denver, the Golden Fault has at least 10,000 feet of vertical movement. Studies of faults and mountain-building uplifts around the world indicate mountains frequently uplift vertically ten or fifteen feet at a time. Rocks involved in the uplift move along faults. Such a movement could create a major *earthquake* such as a 6 or 7 magnitude on the Richter Scale. Assuming the Front Range uplifted ten feet at a time, this would have caused 1,000 major earthquakes! It would appear Colorado has had its share of large earthquakes in the past. On the other hand, if these earthquakes were spread out equally over the past 72 million years of mountain-building, they would have occurred on the average of once every 70,000 years. So when did the last uplift of the Front Range occur? The age of the last movement on a fault may sometimes be ascertained by determining the age of the youngest *sedimentary rock* offset by the fault. On the Golden Fault, the

107

youngest rock (a soil horizon actually) offset has an age of 500,000 years. The amount of fault movement at this time was about ten to fifteen feet. This probably caused a major earthquake. Could the state be overdue for the next one?

An interesting question to ponder is whether the mountains of Colorado have finished uplifting? There is no way geologists can definitively answer the question. However, as long as the North American Continent which is moving west continues to collide with the Pacific plate which is moving east, mountain-building processes will continue in the western U.S. It is possible that Colorado's mountains may uplift again in the future. If you live in Colorado, you might want to check your homeowner's insurance policy for earthquake coverage!

The Front Range contains six "Fourteeners" (mountains over 14,000 ft. in elevation):

Grays Peak:	14,270 ft.
Longs Peak:	14,255 ft.
Mt. Bierstadt:	14,060 ft.
Mt. Evans:	14,264 ft.
Pikes Peak:	14,110 ft.
Torreys Peak:	14,267 ft.

WET MOUNTAINS:

A smaller range southwest of Canon City. The Wet Mountains rise rapidly from the plains to the east. The plains are arid and the mountains receive ample precipitation, hence the name "Wet Mountains. They are composed of Precambrian *granite*. On the northeast side, the granite of the Wet Mountains is faulted against *limestones* of Ordovician, Devonian and Mississippian ages. On the south side, the granite is faulted against Pennsylvanian, Permian and Mesozoic rocks. Three small *stocks* of Cambrian *igneous intrusives* occur within the northwestern portion of the range. Overall, Cambrian *igneous rocks* are uncommon in Colorado.

SPANISH PEAKS:

The Spanish Peaks are composed of Tertiary-age *volcanoes* with radiating dikes. They are surrounded by *sedimentary rocks* of Cretaceous age through which volcanic magma penetrated (See Figure 44). Most of the upper parts of the volcanoes have been eroded away. Perhaps they would be more accurately described as *volcanic necks*.

Figure 44. Spanish Peaks. These two large mountains are middle Tertiary volcanoes located southwest of Walsenberg. They erupted through surrounding sedimentary rocks of Tertiary and Cretaceous age. Erosion has removed much of the upper portion of the volcanoes. Numerous volcanic dikes radiate out from them.

109

SANGRE DE CRISTO RANGE:

These mountains extend from Salida southward into New Mexico. The range is very narrow and high, containing several 14,000 foot peaks as shown below (See also Figure 45).

Blanca Peak:	14,345 ft.
Challenger Point:	14,080 ft.
Crestone Needle:	14,191 ft.
Crestone Peak:	14,294 ft.
Culebra Peak:	14,069 ft.
Ellingwood Point:	14,042 ft.
Kit Carson Mountain:	14,165 ft.
Humboldt Peak:	14,064 ft.
Little Bear Peak:	14,037 ft.
Mount Lindsey:	14,042 ft.

Northernmost End of Range: Composed of Precambrian *igneous* and *metamorphic rocks*.

Majority of Range: The central portion of the range is largely *Permo-Penn sedimentary rocks* (reddish *sandstones*). When the Sun is low in the evening, these mountains exhibit a dark red color. The early Spanish adventurers who entered this portion of Colorado observed the reddish color and gave the range its name, Sangre de Cristo, which in English means "Blood of Christ". Precambrian *igneous* and *metamorphic rocks* frequently are present on the west side of the higher elevations.

Hot Springs: Several hot springs occur along the major bounding *fault* on the west side of the Sangre de Cristo range. These are located east of the village of Villa Grove. Other hot springs exist in the valley a short distance south of Villa Grove. Hot springs also exist near Alamosa.

WESTERN BOUNDARY OF NORTH, MIDDLE AND SOUTH PARKS:

North, Middle and South Parks are intermontane basins located on the west side of the Front Range. On the west side, they are bordered by a long, north-south trending ridge which is subdivided into several mountain ranges.

PARK RANGE:

Park Range is located in the northern portion of the ridge and forms the western boundary of North Park. It is composed of Precambrian *igneous* and *metamorphic rocks*.

RABBIT EARS RANGE:

The Rabbit Ears Range is south of the Park Range and extends eastward, forming the boundary between North and Middle Parks. It consists of Tertiary *volcanics*.

Figure 45. Sangre de Cristo Range. A view from the west of a portion of the northern Sangre De Cristo mountains. In this area, the lower portion of the range is composed of Precambrian gneiss overlain by Ordovician, Devonian and Mississippian sedimentary rocks. The summits are made up of Pennsylvanian Belden and red-colored Minturn rocks. This is a narrow range of mountains which rise abruptly from the eastern side of the San Luis Valley. Several "Fourteeners" are in the range.

111

GORE RANGE:

The next range to the south of Rabbit Ears Range is Gore Range. In its center, the range is composed of Precambrian rocks (1,700 million years old). On the west flank, red *Permo-Penn* rocks are exposed, as can be seen in the Vail Pass area.

TENMILE RANGE:

This range is comprised of Precambrian *metamorphic* and *sedimentary* Paleozoic rocks. Tenmile Range forms the northern part of the western boundary of South Park. It is a high mountain range, containing the following five "Fourteeners":

Mt. Bross:	14,172 ft.
Mt. Cameron:	14,239 ft.
Mt. Democrat:	14,148 ft.
Mt. Lincoln:	14,286 ft.
Quandary Peak:	14,265 ft.

MOSQUITO RANGE:

Consisting of Precambrian *granite* and sedimentary Paleozoic rocks, the Mosquito Range forms the southern part of the western boundary of South Park (See Figure 46). The range contains one "Fourteener":

Mt. Sherman: 14,036 ft.

BUFFALO PEAKS:

Located at the southern end of the Mosquito Range, these peaks are only "Thirteeners" but they are highly visible in the local area. They are capped by the Middle Tertiary Buffalo Peaks *Andesite* which is underlain by *tuffs* and *breccias*. The total thickness of the *volcanic rocks* is greater than 1500 feet. The *volcano* which erupted these rocks appears to have been somewhere to the west in the Sawatch Range area. Volcanic rocks filled a paleovalley underlain by Pennsylvanian rocks. The surrounding higher areas bordering the valley have since eroded away, leaving the volcanic rocks as mountains (See Figure 47). This is a geologic phenomenon known as *"inversion of relief"*. What is low has become high! Therefore the Buffalo Peaks are mountains formed by *erosion*.

Figure 46. Mosquito Range. Aerial view of Mosquito Range, looking to the west. Essentially, this may be seen by travelers on Highway 285 in the vicinity of Fairplay. In the left middle of the photograph, there appears to be a fold (anticline) in lower Paleozoic rocks. However, this is not the case. Rock layers are inclined to the east and a glacial cirque has been incised back into them. This creates an illusion of a fold, when viewed from the east. Photo by T.S. Lovering, no date. USGS Photo Library.

Figure 47. Buffalo Peaks. These peaks are moderate in height, having a maximum elevation of 13,327 feet. However, they are highly visible from much of South Park and from the Arkansas River Valley to the west. They are located at the southern end of the Mosquito Range and form part of the western border of South Park. The peaks are composed of Middle Tertiary extrusive volcanics which filled an ancient paleovalley. Erosion has since removed the surrounding rocks, leaving the volcanics as peaks.

SAWATCH RANGE:

This range borders the Arkansas Valley on the west. Most of the high peaks of the Sawatch Range are composed of Precambrian *metamorphics*. However, Mt. Princeton (See Figure 48), Mt. Antero, Mt. Tabeguache and Mt. Shavano are composed of Tertiary *intrusive igneous rocks* (*granite* and *quartz monzonite*). In Colorado, Tertiary igneous rocks are generally *extrusive volcanics*. These mountains are unusual because they are composed of Tertiary intrusives.

The Sawatch Range contains Colorado's highest mountains:

Collegiate Peaks: The Collegiate Peaks were named after the Ivy League universities:

Mt. Harvard:	14,420 ft.
Mt. Yale:	14,196 ft.
Mt. Columbia:	14,073 ft.
Mt. Princeton:	14,197 ft.
Mt. Oxford:	14,153 ft.

Huron Peak:	14,005 ft.
La Plata Peak:	14,421 ft.
Missouri Mountain:	14,067 ft.

Mount of the Holy Cross: 14,005 ft. The Mount of the Holy Cross is located in the northern end of the Sawatch Range. It was made famous by a photograph taken in 1873 by the legendary photographer, W. H. Jackson, the chief photographer for the Hayden Survey (See Figure 49). A large cross outlined by snow is highly visible on the upper flank of the mountain. The mountain is composed of Precambrian *gneiss*. Large fractures in the rocks have eroded and hold the snow, forming the outline of a cross.

Mt. Antero:	14,269 ft.	
Mt. Belford:	14,197 ft.	
Mt. Elbert:	14,433 ft.	(Colorado's highest peak).
Mt. Massive:	14,421 ft.	
Mt. Shavano:	14,229 ft.	
Mt. Tabeguache:	14,155 ft.	

ELK MOUNTAINS:

Located between Aspen and Crested Butte, these are mountains formed by *thrust-faulting*. Thrust faulting is caused by crustal compression, where rocks are moved up and over adjacent rocks, sliding on a low-angle *fault*. Mountains formed in this manner are uncommon in Colorado, as most mountains within the state were formed by vertical uplifts of crustal blocks. In Colorado, even mountains created by *volcanoes* are more common than are thrust fault mountains.

The Elk Mountains consist of Pennsylvanian and Permian rocks. They have been thrust to the southwest, over-riding younger Jurassic and Cretaceous rocks. These thrust faults along the southwest side of the mountains may be a southeast extension of the tectonism which created the Grand Hogback *monocline* (discussed on page 132).

Figure 48. Mt. Princeton. Located southwest of Buena Vista, Mt. Princeton at 14,197 ft. is one of the "Fourteeners", but not the highest in Colorado. On the other hand, it is one of the most beautiful. The mountain rises majestically from the floor of the Arkansas River Valley, as viewed in this photo from the east on State Highway 285. The vertical relief from the valley floor to the summit is approximately 5,800 feet. The mountain is a *felsic* intrusive igneous rock known as *quartz monzonite* (37 million years old), similar to granite.

116

156. MOUNTAIN OF THE HOLY CROSS.

Figure 49. Mount of the Holy Cross. 14,005 ft. Picture
taken by W.H. Jackson, 1873, northwest of Leadville in
the Sawatch Range. This is the photograph that made
Jackson, official photographer for the Hayden Survey,
famous and also guaranteed everlasting fame for the Mount
of the Holy Cross. The mountain is composed of
Precambrian gneiss and large fractures in the rock hold
snow into the summer, forming the pattern of a cross.
USGS Photo Library.

117

The Elk Mountains are composed of Paleozoic *sedimentary rocks*, including red-colored Permian and Pennsylvanian *sandstones* and *metamorphic marble*. Several Middle Tertiary *intrusive igneous stocks* occur within these mountains. The famous Maroon Bells (See Figure 50) consist of red-colored rocks of the Maroon Formation. Significant marble deposits occur near the town of Marble. This marble (known as Yule marble) is of excellent ornamental-stone quality and was extensively mined at one time (See Figure 51). The Lincoln Memorial and the Tomb of the Unknown Soldier (See Figure 52) are constructed of Yule marble from Colorado. Municipal buildings in both San Francisco and New York City and the Field Building in Chicago also utilized the marble. In Denver, Yule marble was used for several prominent buildings, including the old Main Post Office, Customhouse, City and County Building, Federal Reserve Bank and State Capitol Annex.

Yule Marble is a recrystallized Mississippian Leadville limestone. *Metamorphism* is believed to have occurred during the intrusion of several *igneous* bodies at the time of Cenozoic mountain-building episodes. Pressure from *overthrusting* may also have played a part in the metamorphism. *Limestone* and *dolomite* recrystallize easily under pressure, forming the metamorphic rock, marble. The marble bed outcrops for a distance of 4,000 feet along Yule Creek and on adjacent slopes of Treasury Mountain. It is about 240 feet thick but not all is useable for quarrying.

Marble deposits were discovered in 1873 and mining began in 1875. From 1875 to 1941, a succession of companies operated the quarries. Steep terrain, *landslides*, mudflows and snowslides plagued operations. This, along with the advent of World War II, changing architectural styles, and rising labor and transportation costs caused the quarry to close down in 1941. Due to the demands for steel during the war, equipment was removed and used as scrap iron. After the war, high costs of transportation and changing construction practices discouraged use of Yule marble. Colorado Yule Marble Company reopened the quarry in 1989 and the high-quality marble is being used for monuments and statues.

The Elk Mountains have six "Fourteeners" as shown below. Two of these, North Maroon Peak and Maroon Peak, are the "Maroon Bells":

Capital Peak:	14,130 ft.
Castle Peak:	14,265 ft.
Maroon Peak:	14,156 ft.
North Maroon Peak:	14,014 ft.
Pyramid Peak:	14,018 ft.
Snowmass Mountain:	14,092 ft.

Figure 50. Maroon Bells. The famous Maroon Bells in the Elk Mountains near Aspen. The rocks are composed of the red Maroon Formation. North Maroon Peak (14,014 ft.) is the closest peak shown and Maroon Peak (14,156 ft.) is behind it and to the left. Photo by Bill Chirnside, 1987.

119

Figure 51.   Marble Quarries.   Yule Marble quarry at
Marble on the southwest side of the Elk Mountains. Yule
Marble is metamorphosed Mississippian limestone (or
dolomite). It is believed to have been metamorphosed by
local igneous intrusions.   Photo by J.W. Vanderwilt,
1938. USGS Photo Library.

Figure 52. Tomb of the Unknown Soldier. The tomb was made from a single block of Yule marble from the quarries at Marble, Colorado. In excess of one hundred workers labored many months to quarry the 124 ton block from which the tomb was cut.

WEST ELK MOUNTAINS:

The West Elks are located west of Crested Butte. These mountains are a series of middle Tertiary *igneous* intrusions. They have diameters ranging from two to four kilometers. Igneous *stocks* intruded Cretaceous rocks and each intrusion today forms a mountain. *Landslides* are very common around edges of the West Elk Mountains. One would be wise not to build a home within areas of potential catastrophic earth movements! A very impressive landslide occurred on West Beckwith Mountain in the West Elk Mountains (See Figure 53).

SAN JUAN MOUNTAINS:

The San Juans are the most extensive mountain range in Colorado and occupy a considerable portion of the southwest corner of the state. The mountains consist of three distinctive areas characterized by different rock types. The largest area is covered by *volcanic rocks*. To the southwest of the volcanic area, Precambrian *igneous* and *metamorphic rocks* are exposed. Fringing the igneous and metamorphic rocks are *sedimentary* layered rocks.

The volcanic area comprises the majority of the San Juan Mountains. It is made up largely by of middle Tertiary volcanics which were ejected by a cluster of *volcanoes* and *calderas*. Repeated eruptions created deposits many thousands of feet thick. Some of the eruptions were gigantic. Over a period of time, up to fifteen calderas were formed. Thousands of cubic kilometers of rock and *magma* were blasted into the atmosphere (Steven and Lipman, 1976). As large volumes of magma were ejected, the vacated magma chambers below collapsed inward, creating the calderas.

Much of the volcanic material expelled fell within the immediate area, covering the landscape with great thicknesses of volcanic ash. Some of the eruptions created hot *ash flows*. Hot, molten fragments of ash mixed with extremely hot volcanic gases and the mixture flowed rapidly down the slopes of volcanoes with great velocity. The resulting ash flow deposits are called "*welded tuffs*". The hot ash particles weld together where they are deposited, creating a very hard *volcanic rock*. Undoubtedly, *lahars* (hot, *volcanic ash* mudflows) were common where accumulations of thick deposits of loose fragments of volcanic ash occurred. Lahars form where ash particles are mixed with water from precipitation.

The Silverton, Lake City, and Creede calderas, named after local towns, have long been recognized by geologists. They are some of the largest calderas in the San Juans. The most significant one in the state is the La Garita Caldera, located on the east side of the San Juan Mountains. Calderas are very large volcanoes that ultimately collapsed inward, forming a large depression. They are recognized only by extensive and lengthy geologic field work.

Figure 53. Landslide. Landslide on west slope of West Beckwith Mountain (12,185 ft.), located in the West Elk Mountains, about 17 miles east of Paonia. The landslide traveled from almost the top of the mountain to the bottom of the canyon, a horizontal distance of one and three-quarter miles. Landslides such as this are referred to by scientists as "long runout landslides" which seem to defy the laws of physics. To be safe, how far away from a mountain does one have to be? West Beckwith Mountain is a middle Tertiary felsic igneous rock which intruded the Cretaceous Mesaverde Formation. A number of similar intrusives are in this general area and many of them form the core of mountains. The photo was taken by the USGS geologist, W.T. Lee in 1909, but he referred to the mountain as being "Mt. Lombard." The name was later changed to West Beckwith Mountain. USGS Photo Library.

123

The La Garita Caldera was mapped by Peter Lipman, a USGS geologist, over a period of nearly thirty years. La Garita was formed by an eruption 28 million years ago and the caldera is a twenty by fifty mile oval. The eruption ejected the Fish Canyon Tuff which is thousands of feet thick. The volume of Fish Canyon Tuff is estimated to be 1,200 cubic miles! To date, this is the largest single volcanic eruption recognized by geologists anywhere in the world! Hot *ash flows* spread out over many tens of miles with velocities up to 100 mph. Based upon the estimated volume of ash ejected from the volcano, this eruption was approximately 4,000 times greater than the eruption of Mt. St. Helens.

On the southwest side of the San Juan Mountains and to the northeast of Durango is a high mountainous area of Precambrian rocks. These rocks are composed of *granites* (1,400 and 1,700 million years age), *gneiss* (1,700 to 1,800 million years age) and *metamorphic quartzite*, *phyllite* and *slate* of the Uncompaghre Formation (1,400 to 1,700 million years age).

Fringing the southwest side of the San Juan Mountains are *sedimentary rocks* ranging from Cambrian through Cretaceous. The Cambrian rocks are represented by the Sawatch Quartzite; Ordovician and Silurian age rocks are absent. Devonian rocks are the Dyer and Parting Formations. Next is the Mississippian Leadville Limestone, followed by extensive exposures of the Pennsylvanian Hermosa Group which is equivalent to the Fountain Formation on the east side of the Front Range. The Hermosa contains *conglomerate*, *sandstone*, *shale* and *limestone*. The *nonmarine* Permian Cutler Formation is present, followed by the Triassic Dolores Formation. The Cutler is composed of conglomerate, sandstone and *siltstone*. The Dolores is red conglomerate, sandstone, siltstone and shale. The ever-present Jurassic Morrison Formation is here, along with the Entrada Sandstone. The Cretaceous Dakota, Burro Canyon, Mesaverde and Mancos Formations are exposed in foothills flanking the mountains.

In summary, the San Juans contain a remarkable assemblage of *igneous*, *metamorphic* and *sedimentary rocks* of many types. The majority of the mountains, however, are composed of *volcanic rocks*, primarily volcanic *tuffs*.

During the intense episodes of eruptions, some *volcanic ash* undoubtedly was carried by winds away from the area of the San Juan Mountains. Much of Colorado likely was covered at times by ash falls from the highly explosive events in the San Juans.

The geologic record of volcanism in the San Juan Mountains is as profound as any other volcanic center in the world. If such intense volcanism was still occurring in Colorado, the state would be vastly different from what it is today, and it probably would be uninhabited.

The present landscape of the San Juan Mountains has been created by *weathering* and *erosion* processes acting upon the thick deposits of *tuffs* and volcanic *breccias* and acting upon the Precambrian *crystalline* and Paleozoic *sedimentary rocks*. During the Pleistocene, much of the mountains were covered by *alpine glaciers* which significantly eroded and modified the landscape. The glacial features seen today were created including *U-shaped valleys*, *cirques*, *moraines*, *horns*, *aretes* and *hanging valleys*.

At the present time, *landslides* in many areas are actively eroding away places of high relief. Landslides occur infrequently but when they do occur, they move large volumes of rock and loose material downslope. Landsliding is a major factor in mass-wasting processes and areas of high relief are slowly eroded away by such processes.

Rock glaciers are very common in the San Juan Mountains. They form where *talus* accumulates at the base of steep mountain flanks and slowly move downslope. Some are thought to be cored with ice.

The San Juan Mountains are characterized by spectacular geologic scenery and many geologic features may be observed. (See Figures 54-58). The mountains contain the following thirteen "Fourteeners":

El Diente Peak:	14,159 ft.
Handies Peak:	14,048 ft.
Mt. Eolus:	14,083 ft.
Mt. Sneffels:	14,150 ft.
Mt. Wilson:	14,246 ft.
Redcloud Peak:	14,034 ft.
San Luis Peak:	14,014 ft.
Sunlight Peak:	14,059 ft.
Sunshine Peak:	14,001 ft.
Uncompahgre Peak:	14,309 ft.
Wetterhorn Peak:	14,017 ft.
Wilson Peak:	14,017 ft.
Wimdom Peak:	14,082 ft.

UINTA MOUNTAINS:

The eastern end of the Uinta Mountains extends from Utah into Colorado. This range trends east-west and is a very large, complex *anticline*. It is composed of Mesozoic and Paleozoic rocks with a Precambrian core and was formed by vertical uplift in the Laramide. The Uinta anticline is subdivided into a number of lesser folds, accompanied by some faulting. The smaller anticlines have flat tops with *monoclinal folding* on the flanks. The *monoclines* are highly visible along the southern flank of the Uinta Mountains.

Sedimentary rocks of the Precambrian Uinta Mountain Group are extensively exposed in the higher portions of the Uinta Mountains. Paleozoic and Mesozoic rocks outcrop on the flanks of the mountains.

Figure 54. Uncompahgre Peak . Uncompahgre Peak (14,309 ft) is located east of Ouray on the northern edge of the San Juan Mountains and is composed of Tertiary ash flow tuffs and andesitic lavas. Photo by W.H. Jackson, 1875. USGS Photo Library.

Figure 55. Rock Glacier. Rock glacier in Silver Basin near Silverton. Rocks fall from steep mountainsides and form talus deposits at the base. These accumulations of rock rubble become cored with a mixture of rock and ice and then flow slowly downhill similar to glacier movement. Photo by W. Cross, 1899, USGS Photo Library.

Figure 56. Ouray. Pennsylvanian conglomerate, sandstone and shale of the Hermosa Group are exposed in the canyon walls above the old mining town of Ouray. Ouray is located on the northwest margin of the San Juan Mountains.

128

Figure 57.   Red Mountain: A view of part of Red
Mountain, located on the west side of the Silverton
Caldera.   The red and yellow colors are called
"alteration colors", and are caused by oxidation of iron
in the volcanic rocks from *hydrothermal* fluids which
mineralized them.   This process emplaced veins of metal
sulfides and the Silverton Caldera was mined extensively
during Colorado's mining days.   The very large Idarado
Mine (not shown) is nearby.   Red Mountain can be easily
viewed from the highway from Ouray to Silverton.   With
its bright colors, you can't miss it!

129

Figure 58. Engineer Mountain. The upper photo shows the summit of Engineer Mountain (12,968 ft.). It is comprised of the Pennsylvanian Hermosa Group in the lower part and the Permian Cutler Formation in the upper part. The lower photo is a closeup of the promontory which extends off the east side of the mountain. Limestones, conglomerates, sandstones and shales of the Hermosa are exposed in the cliff face.

130

Rocks of the Precambrian Uinta Mountain Group are unmetamorphosed light to dark red *nonmarine cross-bedded sandstones* that locally contain pebbles (See Figure 59). A few beds of red to gray *shales* are present. These rocks are from 950 million years to 1,400 million years of age and are believed to be greater than 7,000 meters thick. They are very unusual Precambrian rocks as most Precambrian rocks in Colorado are *granites* and *metamorphics*. These may be part of an old Precambrian-age continent which would have been located in the far northwestern part of the state and adjacent parts of Utah and Wyoming. The Uinta Mountain rocks are the only evidence of any continent in Colorado during Precambrian time.

The *clastics* of the Uinta Mountain Group are frequently described as being *metamorphic quartzites*. Examination of these rocks under magnification reveals fractures that break around the sand grains, not through them, therefore they are sandstones and not quartzites. There is some *silica* cement between the grains which gives the rocks a sparkle resembling that of some quartzites, hence the confusion.

Above the Precambrian is the Cambrian Lodore Formation which is composed of a shoreline sandstone and marine glauconitic shale. A large *unconformity* exists on top of the Cambrian, thus Ordovician, Silurian and Devonian rocks are missing (if they were ever deposited here, they were subsequently eroded away). Above the unconformity are marine limestones, shales and some sandstones of the Mississippian Madison, Doughnut and Humbug formations. Above these lie the Pennsylvanian Morgan Formation (*beach* sandstone and offshore shallow marine limestone and shale) and the Weber Sandstone (largely *eolian* sand dunes). The Permian is represented by the Park City Formation (marine limestone, sandstone and shale). Triassic rocks present are the nonmarine Moenkopi (varicolored *shales*, *mudstones*, *siltstones* with thin beds of *gypsum*) and the Chinle (*cross-bedded* sandstones). Mesozoic rocks fringe the uplifted areas of the Uinta Mountains.

Only one Tertiary formation is present, the Bishop Conglomerate. The youngest geologic materials are a number of Quaternary deposits, including *landslides* of Holocene and Pleistocene age. Many of the landslides occur where streams have undercut incompetent shales in the Lodore, Doughnut and Morgan Formations.

Figure 59. Cold Springs Mountain. This mountain is the northerly limb of the Uinta Mountain Range and lies to the north of Brown's Park in the far northwest part of the state.   The mountain is comprised of Precambrian Uinta Mountain Group red pebbly sandstones. On the east end, light colored Paleozoic rocks plunge (dip) east into the subsurface.   These are the Mississippian Madison Limestone, the Pennsylvanian Morgan Formation and the Weber Sandstone. In addition, there are other rocks of Triassic, Jurassic and Cretaceous ages.

INTERMONTANE BASINS

THE PARKS:

North, Middle and South Parks are located on the west side of the Front Range. The parks are basins which contain *sedimentary rocks* of Cenozoic, Mesozoic and Paleozoic ages. The basins formed when the modern day Rocky Mountains began to rise. They are synclines where the rocks were downwarped between adjacent rising uplifts (Tweto, 1980b, p. 133). The synclines were filled with sediments which eroded off the uplifted areas. In some areas, they are bordered by large, vertical *faults* at the borders of the mountains.

Middle Park also contains a mountain of special interest. Wolford Mountain was formed by *thrust faulting* during the *Laramide Orogeny*. It is composed of Precambrian *granites* (1,700 Million years old) thrust on top of Cretaceous Mancos *shales* (See Figure 60). The *thrust fault* is believed to be the northerly extension of the Williams Fork Thrust, which lies to the southeast and borders the west side of the Williams Fork Mountains. The Williams Fork Thrust is one of the few thrust faults in Colorado. As mentioned before, thrust faults, and related mountains formed by them, are uncommon in Colorado. The majority of mountain-building processes in the state involved vertical uplifting.

SAN LUIS VALLEY:

The San Luis Valley is located in southern Colorado between the Sangre de Cristo Range on the east and the San Juan Mountains on the west. It is a very deep basin formed in the middle Tertiary. Largely filled with Tertiary *volcanics* from eruptions in the San Juan Mountains. The valley also contains Cenozoic age *sedimentary conglomerate*, *sandstone* and *shale* derived from *erosion* of the San Juan Mountains and, to a lesser extent, the Sangre de Cristo Mountains.

The San Luis Valley is an extension of the *Rio Grande Rift* which is still active. The Rio Grande Rift follows the Rio Grande Valley from New Mexico and continues northward through the San Luis Valley. A rift begins when crustal spreading starts below a continent and starts to break the continent apart. If the rifting proceeds long enough, it may separate continental rocks far apart so that the ocean waters flow in and a new ocean is created. Then the *spreading center* becomes a mid-oceanic ridge beneath the ocean.

The Rio Grande rift is widening at a rate of one millimeter/year and basaltic magma is moving up and swelling the ground surface in an area near Socorro, New Mexico. The U.S. Geologic Survey has been monitoring this activity. If rifting continues, it may, given enough geologic time (millions of years),

Figure 60. Wolford Mountain. Located in Middle Park north of Kremmling, the upper part of the mountain is composed of Precambrian granite (1,700 million years old) which has been thrust over Cretaceous Pierre shales. The thrust is located at the base of the trees. This thrust is an extension of the Williams Fork Thrust which trends to the southeast. Thrust faults are not common in Colorado.

split portions of the North American continent apart and the Gulf of Mexico would flow in. If this were to happen, New Mexico and Colorado would be divided into halves with an ocean in between. It would require many tens of millions of years for this to happen at the present rate of slow rifting. The rifting may later speed up or could cease entirely. There is no geological guarantee it will continue. If you are a speculative type, buy your beach-front property now!

ARKANSAS RIVER VALLEY: The Arkansas River Valley extends from Leadville on the north to Salida on the south. It is believed to be a northerly extension of the *Rio Grande Rift*.

PLATEAUS:

Westernmost Colorado is a high area of flat-lying *sedimentary rocks* of Paleozoic-Cenozoic ages 5,000 feet or more above sea-level. River drainages have deeply dissected the terrain, creating *plateaus* and *mesas* separated by deep valleys. The rocks exposed at the surface are usually Jurassic in age. A few local *igneous* intrusions and *volcanoes* occur.

WHITE RIVER PLATEAU:

The White River Plateau is located in northwest Colorado and lies north of Interstate 70 between Rifle and Dotsero. It is a high plateau, reaching elevations of 11,000 feet. The White River Plateau is a region which was uplifted vertically during the *Laramide Orogeny*. The western and southwestern sides of the uplift are bounded by erosional remnants of an extensive *monocline*, known as the Grand Hogback. The monocline is easily visible from the highway between Rifle and Meeker and also may be viewed from Interstate 70 between Rifle and Newcastle, where the monocline crosses the interstate and trends to the south along the west side of the Elk Mountains. This monocline is one of the longest geologic structures in the state. Many formations of Cretaceous, Jurassic, Triassic, Permian and Pennsylvanian ages are exposed in the steeply dipping rocks of the monocline (See Figure 61).

The west and southwest areas on the top of the uplift contain exposures of several formations of Pennsylvanian, Mississippian, Devonian, Ordovician and Cambrian ages. These strata are relatively horizontal.

The northeast side of the uplift is covered by Tertiary *basalts*, *tuffs* and *volcanic breccias*. This area is west of Yampa and is known as "The Flattops."

135

Figure 61. Grand Hogback. The remnants of a monocline
which dips west along the west flank of the White River
Uplift can be seen from the highway between Meeker and
Rifle. The sandstones and shales are equivalent to the
Upper Cretaceous Mesaverde Formation. Older rocks (not
shown) in the Grand Hogback are Cretaceous through
Pennsylvanian in age, notably the Mancos, Dakota,
Morrison, Entrada, Glen Canyon, Chinle, State Bridge,
Weber and Maroon Formations. When the White River
Plateau was uplifted, these rocks were folded into a
monocline along the edge of the uplift. Erosion has
since removed them from the top of the plateau, leaving
a series of ridges and *flatirons* composed of steeply
dipping rock.

UNCOMPAHGRE PLATEAU:

The Uncompahgre Plateau is located south of Grand Junction. It is one of the lower plateaus, reaching elevations of 7,500. The Uncompahgre Plateau is composed largely of exposures of Cretaceous, Jurassic and Triassic rocks. In a few places, streams have eroded into *igneous granites* of 1,400 and 1,700 million years age and *metamorphic gneisses*, *schists* and *migmatites* of 1,700 to 1,800 years age. Paleozoic rocks are not present. The plateau is within the area of the ancient Pennsylvanian-age Uncompahgre Uplift where the Paleozoic rocks were eroded away. Therefore, in this area, there is a major *unconformity* composed of Mesozoic rocks on top of Precambrian rocks. Large *monoclines* exist along the northern side of the *plateau* (for more information on these, see the Colorado National Monument section of this guidebook.

ROAN PLATEAU:

The top of the 7,500 ft. Roan Plateau is veneered by Tertiary sandstones and siltstones of the Uinta Formation (do not confuse this formation with the Precambrian Uinta Mountain Group!). The Green River and Wasatch Formations underlie the Uinta Formation. The Green River Formation is well exposed in the cliffs which fringe the plateau. The cliffs along the southeast side of the plateau are called the Roan Cliffs, a portion of which is shown in Figure 39. The Roan Cliffs are visible from I-70 northwest of the Rifle area. The Roan Plateau contains the majority of the *oil shales* discussed earlier.

GRAND PLATEAU:

The 10,500 ft. Grand Plateau may be subdivided into two smaller *mesas*, Battlement Mesa to the north and Grand Mesa to the south. The Grand Plateau was once joined with the Roan Plateau to the north and is composed of the same rocks (Tertiary Uinta, Green River and Wasatch Formations). The Grand Plateau was separated from the Roan Plateau by erosion from the Colorado River. The Grand Plateau also was dissected by Plateau Creek, forming Battlement Mesa and Grand Mesa. Battlement Mesa is capped by Tertiary *basalt lava flows* and flanked by extensive *landslide* deposits. Grand Mesa is also capped by Tertiary basalt flows and is flanked by glacial drift deposits.

MESA VERDE PLATEAU:

The 7,000 to 8,000 foot Mesa Verde Plateau is capped by sandstones and shales of the Cretaceous Mesaverde Group and underlain by *marine* shales of the Mancos Formation. The Mesa Verde Plateau is discussed in detail in the Mesa Verde National Park Section of this guidebook.

EASTERN PLAINS:

SURFICIAL:

Modern soils of Pleistocene and Holocene age overly Tertiary and uppermost Cretaceous rocks in the eastern plains of Colorado.

SUBSURFACE:

Approximately 10,000 to 14,000 feet of *sedimentary rocks* underlie the soils of the eastern plains of Colorado and represents the most complete sedimentary sequence in Colorado. These rocks range from Tertiary through Cambrian age. They overlie Precambrian *igneous* and *metamorphic rocks* which may extend down to the base of the continent. In general, these subsurface rocks are horizontally bedded. Near the mountain fronts, rocks are uplifted and exposed by *erosion*.

VOLCANIC ROCKS:

Some *volcanic rocks* of Tertiary age are exposed near the eastern edges of the mountain fronts.

NORTH AND SOUTH TABLE MOUNTAINS:

These landmarks of the town of Golden are basaltic volcanic *lava flows*. They are believed to have originated from an *igneous* dike to the north. The age of the basalt is 62-64 million years old and is equivalent in age to part of the Denver Formation.

HUERFANO BUTTE:

Just north of Walsenburg, a small remnant of a *volcanic neck* may be seen just east of Interstate 25 (See Figure 62). This extinct *volcano* looks decidedly out of place as it is in an area of *sedimentary rocks*. The volcanic rock is a type of *gabbro*, an *intrusive igneous rock* of *mafic* composition. Gabbro is an uncommon rock in Colorado.

VOLCANIC DIKES:

A large number of Tertiary *volcanic dikes* (vertical sheets of *igneous rock*) are present in the area between Walsenberg and Trinidad both east and west of Interstate 25. These dikes fill fractures in the horizontally bedded Tertiary and Cretaceous *sedimentary rocks*.

The most spectacular dikes are those which radiate from the Spanish Peaks into adjacent sedimentary rocks. Intrusion of the *volcanoes* fractured the rocks and the fractures then filled with *magma*. The dikes may be viewed from the highway between Walsenberg and La Veta Pass, and from the highway between Walsenberg and Cuchara Pass (See Figure 63).

RATON MESA AND MESA DE MAYA:

These two large *mesas* are composed of Tertiary-age *basalt* flows and are located south of Trinidad and east along the Colorado-New Mexico border. Raton Mesa may be seen from essentially anywhere near Trinidad. Mesa de Maya may be viewed from U.S. Highway 160 east of Trinidad.

Figure 62. Huerfano Butte. The Butte is a volcanic neck composed of an igneous gabbro, which is an unusual rock for Colorado. There are many granites but few gabbros in this state. The Butte was formed by erosion of a Tertiary volcano which pierced the Cretaceous sedimentary rocks which underlie this area. Although not large, it is easily visible on the east side of Interstate 25 about ten miles north of Walsenberg.

139

Figure 63. Volcanic Dikes of the Spanish Peaks. Many volcanic dikes radiate out from the Spanish Peaks located southwest of Walsenberg. Cretaceous rocks were fractured as magma in the volcano forced its way upward. Magma from the volcano then squeezed its way into the fractures, creating vertical dikes. This photo shows the "Devil's Staircase", a large dike visible from State Highway 12, southwest of Walsenberg.

Cross-bedded sandstones of the Lyons Sandstone in Boulder Mountain Park west of Boulder. The cross-beds are believed to be the foreset beds of large paleo-sand dunes.

141

COLORADO'S PARKS AND MONUMENTS

ROCKY MOUNTAIN NATIONAL PARK:

"Where Trees Lie Down To Grow"

Rocky Mountain National Park is located in the northern portion of the Rocky Mountains immediately west of Estes Park (See Figure 64). It is the highest part of the Front Range with Long's Peak at a lofty elevation of 14,255 feet. Much of the park is above timberline. W.T. Lee, a USGS geologist, visited the park in 1916 and remarked it was a place where "trees lie down to grow". Indeed, trees at timberline do grow lying down as a consequence of the strong, cold winds of winter. The park is an area of truly superlative scenery and is one of the most important tourist destinations in the state.

The main entrance is located on the west side of Estes Park. From Estes Park, Trail Ridge Road winds westerly across Rocky Mountain National Park, reaching an elevation of 12,183 feet. Trail Ridge Road is the highest paved road in the United States. The road completely traverses the park and ultimately connects with the west park entrance near Grand Lake. Much of the road is above timberline and offers stunning views.

Longs Peak, one of the "fourteeners", is a magnet for hikers and rock climbers (See Figure 65) and Longs Peak has claimed the lives of many people over the years. Hikers and climbers have fallen to their deaths, and hypothermia and lightning are almost daily summer threats. With regard to <u>all</u> of Colorado's high peaks, the casual hiker would be well advised to <u>be off the summit by noon</u> as fierce afternoon storms are routine. High-altitude sickness is also a serious threat to people not acclimated to such high elevations. Symptoms range from a headache to nausea to pulmonary edema. The headache usually can be cured by an aspirin. Pulmonary edema consists of a buildup of fluid in the lungs, which can on occasion be fatal. In the case of pulmonary edema, the victim should get out of the high country to a lower elevation as soon as possible. "Low-landers" commonly experience shortness of breath at such elevations and this is nothing to worry about unless one has lung or heart problems. Casual visitors should avoid straining their lungs while hiking as this could bring on pulmonary edema.

GEOLOGY:

The majority of the rocks exposed in the mountains of the park are *metamorphic* and *igneous* of Precambrian age. The metamorphic rocks are largely *gneisses* with lesser amounts of *schist* and *quartzite*. The igneous rocks are *batholiths* and *stocks* of *granite*. *Granitic pegmatites* are also abundant in the metamorphic rocks.

Volcanic rocks of Tertiary age are located in the Never Summer Range on the western border of the Park and in the northwest corner of the park. They are flow rocks composed primarily of *basalt* and *andesite*. Lesser amounts of explosive volcanic deposits including ash, *tuff*, *welded tuff*, *breccias*, *lahars* and even some *obsidian* are also present.

Metasedimentary rocks of Cretaceous age are sparingly and surprisingly present in the park. The mountains in the park have undergone incredible *weathering* and *erosion* which have largely stripped them of the vast majority of sedimentary rocks which once covered the Precambrian *basement rocks*. Small patches of metamorphosed shales (*hornfels*) of the Cretaceous Pierre Formation are located in the Never Summer Mountains. These rocks were heated by Tertiary *volcanic* activity and metamorphosed into a durable hornfels which has resisted erosion. The hornfels still contain typical marine fossils of the Pierre such as *Baculites* (an ammonite) and *Inoceramus* (a large clam).

Figure 64. Estes Park. Main street of Estes Park in 1912. Photo by R.B. Marshall, USGS Chief Geographer. USGS Photo Library.

Figure 65. Longs Peak. Longs Peak (14,255 ft) and Chasm Lake. Photo by W.T. Lee, 1916. USGS Photo Library.

ORIGIN OF THE PARK:

The mountains in Rocky Mountain National Park were created during the *Laramide Orogeny*, a major mountain-building episode, beginning about 72 *M.Y.A.* These mountains were largely uplifted vertically many thousands of feet by compression created from the collision of the North American Plate which was moving to the west with the Pacific Plate which was moving to the east. The building of these mountains was episodic. There were periods of uplift followed by periods of *weathering* and *erosion* which reduced the height of the mountains. These events produced relatively smooth surfaces of erosion, some of which may be seen capping some of the high mountains. The last major stage of uplift during the Pliocene raised them to their present lofty heights.

As long as the North American Plate pushes against the Pacific Plate, the possibility of future uplifts of the Rocky Mountains exists. Such an uplift would create many *earthquakes* and explosive volcanism is not out of the question. At the present time, there is no geologic evidence to suggest future uplifting.

Since the Pliocene, the processes of weathering and erosion have sculpted the rocks into the mountains and valleys of today. The most important agent of weathering and erosion has been *glaciation*.

GLACIAL FEATURES:

The most spectacular geologic process, which may be seen today in Rocky Mountain National Park, is the result of glaciation. About two million years ago, the earth cooled and extensive glaciation of the Pleistocene Epoch began. Alpine glaciers formed in the high areas of the park. These sculpted the mountains into a variety of glacial landforms. Glaciers are highly erosive, and of all erosional geologic processes, glaciation is the most efficient. It rips out hard rock and carries it away at a very high rate relative to geologic time. No other erosional process works as fast. It is fortunate that widespread glaciation is not an abundant process through time, else the earth would have few high mountainous areas left!

During the Pleistocene, numerous and large alpine *valley glaciers*, generally located on the northeast side of the mountains, eroded deep *U-shaped valleys* and created numerous glacial features. *Glaciation* indeed is largely responsible for the high relief in the park and for much of the majestic scenery. A number of small glaciers still exist (See Figure 66). The following glacial features may be observed in the park:

cirques	*polished glacial pavement*
U-shaped valleys	*rock glaciers*

145

moraines
   lateral
    terminal
    erratics
paternoster lakes
roches montonnees
striations

felsenmeer
patterned ground
permafrost (not visible)
solifluction
permanent snowfields
small *glaciers*

The Pleistocene glaciation in the park ended 12,000 years ago. The climate of the earth then became warmer and non-glacial conditions returned to most of the Park.

Figure 66. Hallet Glacier. Old crevassed ice is exposed at top center and center left. The *crevasses* are proof this is a glacier which slowly moves downhill and not just a permanent snow field. Crevasses are formed by the movement of glacial ice over uneven terrain. Photo by W.T. Lee, 1916. USGS Photo Library.

MESA VERDE NATIONAL PARK:

Mesa Verde National Park was created in 1906 specifically to preserve *Anasazi* archeological sites, many of which are outstanding examples of cliff dwellings. The Anasazi culture has been extensively studied and many publications are available. Apart from the archeological sites, Mesa Verde is very scenic and laced with spectacular canyons.

GEOLOGY:

Mesa Verde is a *cuesta*, that is to say, a *mesa* whose surface is gently inclined in one direction. Think of a flat-topped table and tilt it slightly in one direction, and you have a cuesta. The Mesa Verde cuesta originally was part of a larger *pediment* that extended to the San Juan Mountains. Rivers and *weathering* processes removed much of the pediment and Mesa Verde is an erosional remnant. Tributaries of the Mancos River have cut steep-walled canyons back into the cuesta. This created excellent sites for the Anasazi to build dwellings in *alcoves* along the cliff faces.

The geology of Mesa Verde is simple. The oldest rocks exposed within the park are shales of the Cretaceous Mancos Formation. Above these are rocks of the Mesaverde Group, principally sandstones and shales. The Mesaverde Group was in fact named for the rocks in the park (yes, Mesaverde as used in the group name is one word!). The youngest geologic materials are unconsolidated Cenozoic-age *pediment* gravels, stream deposits, *talus* and soils.

MANCOS SHALE:

During this time period, *marine shales*, *mudstones* and fine-grained *sandstones* were deposited in the Cretaceous Sea, which by this time had covered southwestern Colorado. The Mancos deposits are easily erodible and forms slopes and valleys.

MESAVERDE GROUP:

The rocks of the Mesaverde Group were deposited as the ocean began to retreat to the northeast. The Mesaverde is composed of shoreline deposits of *sandstones* and *shales* (See Figure 67).

Point Lookout Sandstone:

During this depositional cycle, massive fine-grained and *cross-bedded* sandstones containing *marine* fossils were deposited. The fossils indicate deposition in shallow oceanic waters affected by waves and currents. These rocks are resistant to *weathering* and form cliffs.

Menefee Formation:

Rocks of this formation are characterized by interbedded carbonaceous *shales*, *siltstones* and *sandstones* that formed in a *marginal marine* environment of *swamps* and *lagoons* landward of the shoreline. A few thin coal beds are present. These rocks form slopes between the Point Lookout and Cliff House sandstones.

Cliff House Sandstone:

Rocks deposited during this period are *sandstones* that are massively bedded, fine-grained, *cross-bedded* and contain *marine* fossils. A few thin *shale* beds occur. These rocks are resistant to *weathering* and are cliff formers.

ALCOVES:

Alcoves are shallow cave-like indentations in the face of the cliffs. With an overhanging sandstone roof, they were ideal places for the *Anasazi* to build their dwellings (See Figures 68 and 69).

Alcoves develop where massive sandstones overlie shale beds. The sandstones of the Mesaverde are *porous* and *permeable*. Precipitation on the top of the *mesa* percolates into the sandstone and moves downward as ground water until it encounters an *impermeable* shale bed. Here it is forced to move laterally until it seeps out onto the face of a canyon wall. Sandstones in the area of the seep remain constantly damp and are subject to chemical *weathering* processes (which weakens the cement that holds the sand grains together). Freezing and thawing additionally weaken the cement, and the sandstone erodes away grain by grain, creating the alcoves.

Figure 67. Sandstones of Mesa Verde. A view of the west side of the canyon opposite Cliff Palace. A thick sandstone wall rises form the bottom of the canyon. Above it is a slope with Juniper trees. Above the slope is an upper sandstone cliff, the lower portion of which contains the Cliff Palace (not shown). In the photograph, the lower sandstone is the Point Lookout Sandstone. The slope above is the Menefee Formation. The upper sandstone cliff is the Cliff House Sandstone. The Cliff House Sandstone contains most of the alcoves and cliff dwellings.

Figure 68. Cliff Palace. Cliff Palace viewed from the west. This picture shows an excellent view of the alcove within which the Anasazi built their dwellings. It is easy to see how the alcoves provided shelter from the elements and also provided a defensible position.

Figure 69. Oak Tree House. One of the many cliff dwellings situated in canyon alcoves in Mesa Verde National Park. All suitable alcoves were utilized by the Anasazi.

BLACK CANYON OF THE GUNNISON
NATIONAL MONUMENT:

Black Canyon of the Gunnison (See Figure 70) is located in west central Colorado. The canyon is a 2,700 foot deep chasm cut by the Gunnison River during the Late Tertiary. In places, the canyon is deeper than it is wide, narrowing to 1,100 feet across. Needless to say, it is awe-inspiring with nearly vertical walls extending precipitously down to the river. The walls are composed of Precambrian *gneisses* and *quartz-mica schists*. These rocks have been intruded by many *granitic pegmatites*, creating a complex mosaic of light-colored rock bands. The pegmatites are interbedded with the banding of the *metamorphic rocks* in some areas (See Figure 71). In other areas, they cut across the dark metamorphic rocks. This canyon attests to the erosive power of a river with a steep *gradient*.

COLORADO NATIONAL MONUMENT:

West of Grand Junction and south of Interstate 70, lies the Colorado National Monument. It is located on the northern end of the Uncompaghre Plateau. The *plateau* was uplifted during the *Laramide Orogeny* of Late Cretaceous and Early Tertiary, creating a large, easily visible monoclinal *fold* in the rocks on the northeastern side (See Figure 72). The Monument is an area of beautiful scenery composed of impressive steep-walled canyons eroded into the *plateau*. These canyons are the reason for the designation of the area as a Monument.

Rocks exposed in the canyon walls are of Triassic age. Jurassic and Cretaceous rocks are present farther back from the canyons. These are *sedimentary rocks* composed of *shales* and *sandstones*. At the base of the canyon walls, Triassic rocks overlie Precambrian *metamorphic gneisses* and *schists*. This is a major *unconformity* and was created by *erosion* of the ancestral Uncompaghre Uplift of Pennsylvanian age. The ancestral mountains in this area were not completely eroded away, and covered by younger sediments, until the Triassic Period. The Triassic Chinle Formation overlies the Precambrian, and the Triassic Wingate Sandstone forms the cliffs above (See Figure 73). Erosion of the sandstones within the canyons has produced scenic *pinnacles* (See Figure 74). Receding back from the cliffs are the Triassic Kayenta Formation, the Jurassic Entrada and Morrison formations and the Cretaceous Burro Canyon and Dakota formations. Remnants of Cretaceous Mancos Shale cap the centers of the *plateaus*.

152

Figure 70. Black Canyon of the Gunnison. This picture, taken from Oak Flat Trail at the Gunnison Point Visitor Center, shows the rugged canyon. The Gunnison River is visible at the bottom in the center of the photo. Crystalline gneisses in the canyon walls are hard and resistant to erosion, creating the steep sides. If these rocks were soft and non-resistant to erosion, this would be a wide valley instead of a deep chasm.

153

Figure 71.   Metamorphics of the Black Canyon of the Gunnison. Precambrian gneisses are well displayed in the near vertical canyon walls at Gunnison Point. The light-colored layers are pegmatites which were injected into the gneiss. Here, the pegmatite veins are parallel to the layering of the gneiss. At other places in the canyon, the pegmatite veins cut across layers of the gneiss.

154

Figure 72.   Breached Monocline.   A view of a large monocline dissected by erosion in Colorado National Monument.   It trends along the northeast flank of the Uncompaghre Plateau.   Wingate Formation red sandstones overlie the Chinle with Precambrian granite at the base. Photo by W.T. Lee, 1925.   USGS Photo Library.

Figure 73. Colorado National Monument. Red Canyon,
looking east. Triassic Wingate Formation sandstones form
the massive cliffs and are underlain by the Triassic
Chinle Formation (red shales). Grey colored Precambrian
gneisses are locally exposed in the canyon bottom.

Figure 74. Pinnacles. Sandstone pinnacles formed by weathering and erosion in Monument Canyon of the Colorado National Monument. These pinnacles are remnants of the Triassic Wingate Sandstone.

157

DINOSAUR NATIONAL MONUMENT:

Dinosaur National Monument, which straddles the Utah-Colorado border, is located along the southern boundary of the Uinta Mountains where the Jurassic Morrison Formation is exposed. The Monument represents one of the most famous and important dinosaur *fossil* discoveries in the western United States. Even more importantly, visitors may view a large number of dinosaur bones in place in the rock where they were discovered (See Figure 75). This rock was once a large sandbar in a stream on the Morrison landscape. Dinosaurs did not live in streams so it is a puzzle why so many bones are preserved in what is obviously a sandbar. Presumably, dinosaurs occasionally drowned at stream crossings and bodies would float downstream and lodge on sand bars. Scavengers likely ate the flesh and the bones were incorporated in the sediments comprising the bar.

The initial discovery was made by the Carnegie Museum in 1909. Field crews excavated at the site for 13 years and shipped several hundred tons of bones to the Museum. Many skeletons have been assembled from the bones, representing ten different species of dinosaurs.

The dinosaur bone quarry is located in the Utah portion of the Monument. In the Colorado portion, part of the monument extends to the north and includes the confluence of the Green and Yampa rivers and the magnificent canyons carved by them.

Rocks in the canyons and adjacent highlands vary in age from Precambrian to Quaternary (except for those of Ordovician, Silurian and Devonian ages, which are not present). There are few roads in the Monument and some of the most beautiful canyons can only be viewed by rafting. Harpers Corner, which can be reached by a paved road, offers stunning views of the Green and Yampa river canyons in the area of the confluence.

In the canyon walls of the Yampa River, the Pennsylvanian Weber Sandstone is exposed. On the north side of the river, the Weber is 300 meters thick, and many hundreds of meters of highly *cross-bedded eolian sandstone* are displayed in a sweeping panorama (See Figure 76). A large *monocline* known as the Warm Springs Monocline trends easterly north of the Yampa. Sandstone beds flex downward from the monocline toward the river. On the south side of the river, upper parts of the canyon walls are capped by the younger Permian Park City Formation, a *marine fossiliferous limestone*.

Figure 75 Dinosaur Bone Quarry. Two views of dinosaur bones exposed on the quarry face at the western headquarters of Dinosaur National Monument. The quarry contains many dinosaur bones which have been carefully uncovered by workers. These bones are found in a sandstone which was deposited in an ancient river that flowed through the landscape of the Jurassic Morrison Formation. Dinosaurs sometimes drowned while fording the river and their bones were preserved in sands of the river bed. Such occurrences are evidence that dinosaurs may have migrated in herds, much as the wildebeest in Africa do today.

159

Whirlpool Canyon of the Green River may be viewed to the west from Harpers Corner (See Figure 77). Rocks visible in the northern canyon wall are older. They consist of the Cambrian Lodore Formation at the base, overlain by the Mississippian Madison, Doughnut and Humbug formations, followed by the Pennsylvanian Round Valley Limestone and Morgan Formation.

The northernmost reaches of the Green River within the Monument may only be viewed by rafting. The same formations mentioned above outcrop in the canyon walls, along with the still older Precambrian Uinta Mountain Group. The famous Gates of Lodore, where the Green enters the Canyon of Lodore as it exits Brown's Park, are composed of the Uinta Mountain Group sandstones.

Figure 76. View from Harper's Corner. In Dinosaur National Monument, breath-taking panoramas may be seen from Harper's Corner. The Yampa River is visible on the right side and Green River at the base of the photo. The confluence of the Green and Yampa rivers occurs out of view behind Steamboat Rock in the foreground. Many of the rocks visible in the photo are sandstones of the Pennsylvanian Weber Sandstone. These beds are the southward-dipping flank of the Warm Springs Monocline, whose top is out of the photo to the left.

160

Figure 77. Whirlpool Canyon. Whirlpool Canyon and the Green River as seen from Harper's Corner overlook in Dinosaur National Monument. At the base of the canyon wall, sandstones of the Cambrian Lodore Formation are exposed. Unconformably overlying the Lodore are Mississippian and Pennsylvanian limestones and shales.

161

FLORISSANT FOSSIL BEDS
NATIONAL MONUMENT

The National Park System created the Florissant Fossil Beds National Monument in 1969 to protect fossil-bearing beds of unusual significance. The Florissant Lake Beds have a total outcrop area of about 15 square miles which lie in a narrow basin about one mile wide and 15 miles long. Lake Florissant extended from near Lake George along Highway 24 to the village of Florissant where it turned south and continued for another 7 miles. The Monument contains about 6,000 acres. The lake beds not included in the Monument are on private property.

Fossils found in the lake beds include the petrified stumps of giant redwoods in growth position (See Figure 78), a myriad assemblage of delicate insect fossils (See Figure 79) and abundant tree leaf fossils. The fossil redwoods are spectacular but the insect and tree leaf fossils are of greater scientific importance. Fragile insects are seldom preserved as fossils. Those at Florissant yield very valuable information on the evolution of insects. Tree leaves are uncommonly fossilized (as they are very fragile), and when fossilized can yield valuable knowledge.

The Florissant Fossil Beds are located south of the town of Florissant on U.S. Highway 24 west of Colorado Springs. The spectacular fossil redwood stumps may be viewed at the Visitor Center in the Park. Within the Center, an excellent display of insect and leaf fossils may be seen.

GEOLOGY:

The hills around Florissant are composed of Pikes Peak Granite of Precambrian age (one billion years old). These *granites* are the base below the fossil-bearing beds. The beds are largely *shales* of very latest Eocene age (slightly more than 34 MYA). During this time, the modern-day Rocky Mountains were uplifting and there were many active *volcanoes* in western portions of the state. Repeated eruptions showered ash on the Florissant area. Some *ash flow* events (*nuee ardentes*) created "*welded tuffs*", a hard *rhyolite* formed from hot, molten ash particles which fell and "welded" together. Ash flows are surges of superheated air and molten ash fragments which flow rapidly down the slopes of volcanoes and may travel for many miles. Additionally, tremendous volcanic mudflows (*lahars*) came from the west, and covered parts of the land.

162

Figure 78.   Fossil Redwood.   A number of very large petrified redwood tree stumps in growth position may be viewed at Florissant Fossil Beds National Monument. Redwoods grew in the area before the lake was formed by a volcanic lahar which came from the west.   These trees are thought to be a forerunner of the modern redwoods (Sequoia sempervirens) which grow today in the fog belt of the coast ranges of California.   Modern redwoods commonly reach heights of 200-275 feet with diameters of 8-12 feet.   The diameters of the petrified stumps are within this range, suggesting some rather tall trees in Colorado during this time (Preston, 1961).   Photo courtesy of Beth Simmons.

163

Figure 79. Fossil Insect: Soldier beetle (*chauliognathus*) from the Florissant Fossil beds. Identified by Rick Piegler of the Denver Museum of Natural History. The body of the beetle is one inch long. Simmons Collection.

The source of the various Tertiary volcanic deposits in the Florissant area is not known with any degree of certainty. Welded tuffs may have come from sources in the Sawatch Range west of the Arkansas River Valley. A large volcanic area (the Thirty-Nine Mile volcanic field) was located about 16 miles to the southwest of the Monument and it may have provided some of the *andesitic* lahars which blocked the drainage and created the lake (Epis, et. al., 1980, p. 142).

During this time, large groves of redwoods grew in the paleovalley which now is the Monument. A large *volcanic* mudflow came down the paleodrainage and buried the valley. The redwoods were buried up to a height of 15 feet. The portion of the trees covered by mud was replaced by *silica* from fluids contained in the mud. Silica by itself has an extremely low solubility in water, therefore the source of silica was possibly instead from *silicic acid* which forms from the chemical decomposition of abundant *feldspar minerals* in *volcanic tuffs* (Bloom, 1998). A lake formed and clay, silt and *volcanic ash* fragments were deposited on the lake bottom, creating the shales which today contain the insect and leaf fossils. For delicate insects and leaves to be fossilized, they must be immediately buried in fine-grained sediment (such as clay). The bottom of the lake provided just such a *depositional environment*. Insects and tree leaves occasionally fell into the lake or were carried there by streams. They were buried by clay rapidly enough to be preserved intact.

It is fortunate the fossil beds of ancient Lake Florissant were not later removed by *weathering* and *erosion;* but instead were preserved for our amazement and education.

Fossil collecting within the Monument is prohibited by law, however there is a privately-owned quarry where visitors may collect for a reasonable fee. This quarry is located immediately south of the village of Florissant on the road to the Monument.

LATE EOCENE LANDSCAPE OF COLORADO:

At the time of the growth of the redwoods at Florissant, much of Colorado was near sea level with the early rising Rocky Mountains in the western portion of the state. Many of these early mountains were explosive *volcanoes* and they repeatedly erupted over a time span of millions of years. Some of these were giant eruptions, creating *calderas*.

According to MacGinite (1953, p.57), the climate, based on the types of trees and plants that grew in the area was warm temperate. The average annual temperature was at least 65°F and freezing temperatures were uncommon. The annual rainfall was between 20 to 25 inches with evidence of a pronounced dry season. The area of Lake Florissant had an elevation of about 3,000 feet and the volcanic mountains to the west had elevations 8,000 feet or more.

Large redwood groves grew all over Colorado, as we find pieces of fossilized redwood all over the state. Indeed, these redwood groves grew over much of the Rocky Mountain region.

With the combination of scenic *volcanoes* and groves of giant redwoods growing seemingly everywhere, Colorado must have been breathtakingly beautiful, perhaps even more beautiful than the high alpine Rocky Mountains of today. Unfortunately, such vistas are only seen in the minds of geologists. After all, the rocks do speak to geologists!

GREAT SAND DUNES NATIONAL MONUMENT

Great Sand Dunes National Monument is located in southern Colorado northeast of Alamosa on the east side of the San Luis Valley. The sand dunes are very large, some as high as 700 feet above surrounding areas (See Figure 80). The greatest accumulation of sand is in an area where Medano Creek emerges from Medano Pass in the Sangre de Cristo Mountains north of Sierra Blanca. Medano Pass is a low pass (10,150 ft.) in the Sangres. The Sangre de Cristo Mountain Range is high and contains several 14,000 ft. peaks.

It is surprising to many (including geologists) that such an extensive accumulation of large sand dunes exists in an *intermontane* basin within a state more aptly characterized by its alpine nature. The origin of the sand dunes is controlled by a combination of geographic, geologic and climatic factors.

The San Luis Valley lies between the San Juan Mountains to the west and the Sangre de Cristo Mountains to the east. The valley is 105 miles long and 40 miles wide. It terminates to the north at Poncha Pass that separates the Rio Grande drainage basin from that of the Arkansas River. To the south, the valley extends a short distance into New Mexico.

In the valley, the principal stream is the Rio Grande River which enters the valley from the west (east side of the San Juan Mountains). The river flows southeasterly through the San Luis Valley. North of Alamosa is an area of interior drainage where small streams draining from the Sangres sink into sediments of the valley.

Figure 80. Sand Dunes. A late afternoon view to the
north of dunes of the Great Sand Dunes National Monument.
Prevailing winds are from the southwest. In the picture,
the wind direction is from left to right. The slip faces
(steep slopes) of the dunes are shadowed by the low angle
of the sunlight.

The San Luis Valley is arid with less than 10 inches annual precipitation. Evaporation and transpiration are high. Strong winds blow throughout the year from the southwest. Occasionally, southwest winds are so strong that dust storms are created. Clouds of dust are lifted across the mountains to the east. During the winter, storms sometimes have high-velocity cold winds that blow from the northeast for short periods of time, a reversal of direction.

GEOLOGY:

Approximately 52% of sand grains in the dunes are composed of small fragments of *rhyolite*. 29% of the grains are *quartz*. The remainder are other *minerals*. Average size of sand grains is 0.2 to 0.3 millimeters. Grains up to 2 millimeters occur occasionally in the higher velocity areas of the dunes (See Johnson, 1968).

Geologic structure and rock types of the San Luis Valley, the Sangre de Cristo Mountains and the San Juan Mountains have largely controlled the formation and composition of the sand dunes.

The San Luis Valley is a structural basin bounded by large, *high-angle faults*. It is filled with many thousands of feet of *alluvial fan* gravel (sediments composed of *volcanic* material and interbedded basaltic *lava flows*). This valley fill is of Tertiary Pliocene and Pleistocene age. It is divided into two formations: the older Santa Fe Formation and the younger Alamosa Formation. Surficial deposits of the valley are Quaternary in age. They include stream deposits, *pediment* gravels, alluvial fan materials and fine-grained playa sediments. Hot springs occur in the valley and are associated with vertical faults. Some recent faults on the eastern side of the valley cut very young Pleistocene or Holocene sediments.

To the east, the Sangre de Cristo Mountains are composed of *igneous*, *metamorphic* and *sedimentary rocks*. Igneous rocks are Precambrian *granite* and *granodiorite*. A small number of lesser Tertiary *granitic intrusives* occur within these mountains. *Metamorphic rocks* are Precambrian *gneisses*. All igneous rocks are a source of *quartz* from *weathering* processes. Sedimentary rocks are Paleozoic and Mesozoic *shale*, *sandstone*, *conglomerate* and *limestone*.

In the vicinity of Great Sand Dunes National Monument, the western slope of the Sangre De Cristo Mountains is comprised mostly of Precambrian granite, granodiorite and gneiss. Most debris entering the valley from the east comes from these rocks. In places, *alluvial fans* extend westward into the valley.

To the west, the San Juan Mountains are composed primarily of *extrusive volcanic rocks* of middle to late Tertiary age. They are largely rhyolitic in composition. These volcanics are a major source of sand grains released by weathering processes. Sand grains from the volcanics are *quartz* and *rhyolite*.

SOURCE OF DUNE SAND GRAINS:

The majority of sand grains which make up the sand dunes come from weathering of the Tertiary volcanic rocks of the San Juan Mountains. The Rio Grande River and its tributaries carried these sand grains (quartz and rhyolite) into the San Luis Valley. A lesser amount of quartz sand comes from weathering of the Precambrian granites, granodiorites and gneisses of the Sangre de Cristo Mountains to the east.

Mass wasting, streams and the formation of alluvial fans move quartz grains into the valley. The Rio Grande transports, reworks and deposits sands. Strong winds from the southwest erode sand from valley sediments and move it to the sand dunes. Winds have to rise to get over the Sangre de Cristo Mountains. They are channeled into the area of Medano Pass. As the air rises to go over the pass, loss of velocity causes deposition of the sand, forming the dunes.

North of Monte Vista and Alamosa, there are ancient levees and dry oxbow lakes created by the Rio Grande. When the wind blows, sand is still being derived from these deposits, and the dunes are still growing.

DUNE TYPES:

Wind velocity, sand supply and vegetation control formation of the various types of dunes described below.

Transverse Dunes: Very large. These dunes made the Great Sand Dunes famous. Relative stable. High sand supply. Dunes are primarily constructed by winds from the southwest. Winter winds from the northeast sometimes reverse crests of the dunes (See Figure 81).

Longitudinal Dunes: Very large. High sand supply. Wind direction varies slightly, creating dunes with a long axis parallel to the wind direction.

Barchan Dunes: Crescent shaped with horns pointing downwind. Usually separate from each other and move across a bare surface. These form where the sand supply is limited (See Figure 82).

Parabolic Dunes: Commonly form around a blowout. Crescent shaped. Horns point upwind. Apt to be anchored by vegetation. Abundant sand supply (See Figure 83).

169

Climbing Dunes: No regular features. Consist of irregular mounds and swales on steeply rising slopes. Are piled against steep western slopes of the Sangre de Cristo Mountains (See Figure 83).

Blowouts: Areas of *deflation* from wind erosion. Supply sand to dunes.

Figure 81. Transverse Dunes. Long sand ridges in the photo are transverse dunes. These are the largest and most abundant dunes in the Great Sand Dunes National Monument. Along the crests of these dunes may be seen small reversed dunes. Winter winds from the northeast reverse the crests of the larger transverse dunes, which were formed by winds blowing from the southwest.

Figure 82. Barchan Dunes. Small barchan dunes may be
seen just this side of Medano Creek in Great Sand Dunes
National Monument. Wind direction is from the right to
the left side of the photo. Horns of the barchans point
downwind, to the left. The very tiny dots on the dunes
are people.

Figure 83. Parabolic and Climbing Dunes. During high wind storms from the southwest, some sand is blown across to the east side of Medano Creek. This sand forms parabolic dunes and climbing dunes. In the center of the photo is a parabolic dune with the horns pointing upwind (to the right). On the far left side of the photo, small climbing dunes are progressing up the flank of the mountains and have successfully moved out of Great Sand Dunes National Monument.

172

DUNE FIELD STABILITY:

The dominate movement of the dunes is to the northeast. Winter winds, however, frequently blow from the northeast and move sand in the dunes back to the southwest. This acts to stabilize the dunes to some extent.

Medano Creek, which flows westerly out of Medano Pass in the Sangre de Cristo Mountains, erodes away advancing dunes on the east side of the Monument and carries the sand back to the southwest side where winds in turn move it back up onto the dunes. Thus the dunes have a natural sand recycling system! (See Figure 84).

A few inches below the surface, sand in the dunes is wet. As wind erodes the surface of the dunes, erosion is effectively stopped when wet sand is exposed. This greatly inhibits migration of the dunes.

Figure 84. Medano Creek. In the foreground, Medano Creek, flowing from the right to the left, is a very shallow stream which moves great volumes of sand from the east side of Great Sand Dunes National Monument back around to the southwest, where it deposits it. Wind then blows it back onto the dunes.

GARDEN OF THE GODS:

Garden of the Gods, a Colorado Springs city park, contains magnificent *pinnacles* of red sandstone which stand almost vertical. It is truly awe-inspiring to view this surreal landscape (See Figure 85). Many of the *hogbacks* and *flatirons* up and down the eastern flank of the Front Range are composed of rocks of the Fountain Formation. But in the area of the Garden of the Gods, the Fountain is weakly cemented and erodes down to lower *landforms*. The adjacent red sandstones of the Lyons Formation are better cemented and more resistant to *erosion* than the rocks of the Fountain. Consequently, Lyons sandstones form a majority of the pinnacles. It should be noted that a few of the smaller pinnacles are composed of Fountain sandstones.

Balanced Rock, one of the popular attractions, is comprised of Fountain sandstone (See Figure 86). Near Balanced Rock, rare *paleomudcracks* may be observed on the surface of a sandstone bed (See Figure 87).

On the west side of the Garden of the Gods, a major vertical to high-angle reverse *fault*, known as the Rampart Fault, occurs within the Fountain Formation. Total maximum vertical movement on the fault is estimated to be 15,000 feet! This fault, and others like it, are present along the eastern side of the Front Range. They formed during the uplift of the modern-day Rocky Mountains, beginning approximately 72 million years ago. The total movement was accomplished by a series of smaller movements, spread out over millions of years. Of course, every time the fault moved, there would have been a major *earthquake*!

Figure 85. Garden of the Gods. A view of the cathedral spires in Garden of the Gods. They are composed of sandstones of the Lyons Formation. It is a common misconception they are Fountain sandstones. In this area, the Fountain is not well cemented while the sandstones of the Lyons Formation are. Consequently, the Fountain sandstones have eroded away while the Lyons sandstones resist erosion and form the spires.

175

Figure 86. Balanced Rock. Located in the southwest side of Garden of the Gods, Balanced Rock is Fountain sandstone whereas the majority of pinnacles are Lyons sandstone. The Fountain generally contains larger pebbles than the Lyons. Some of these are visible in the sandstone in the foreground of the picture.

176

Figure 87. Paleomudcracks in the Fountain Formation:
Ancient mudcracks located near the Balanced Rock in the
Garden of the Gods. They are clearly outlined by white
sandstone which filled the cracks shortly after they were
formed. These attest to the nonmarine nature of the
Fountain Formation. Additional information on mudcracks
is provided on page 56.

RED ROCKS PARK OF DENVER:

Red Rocks Park, a Denver city park, is located north of Morrison. The park is an area of outstanding geology. The red rocks of the park are the Fountain Formation. Many geologic features may be observed in the various rock outcrops both in the park and adjacent to it. A considerable portion of the geologic history of the State of Colorado may be viewed in the general area of Red Rocks Park.

Rocks ranging in age from Precambrian to Early Tertiary are present. On the west side of the park, *metamorphic* Precambrian rocks (*gneiss*) belonging to the Idaho Springs Formation are exposed. *Sandstones* and *conglomerates* of the Middle Pennsylvanian Fountain Formation overly them unconformably. This geologic contact is a major *unconformity* representing a minimum of 270 million years. Rocks which are missing are approximately one to two miles in thickness. Geologists regard this as evidence that an ancient mountain range rose up and the strata which are no longer present eroded away from their tops. Throughout Red Rocks Park, red *nonmarine* sandstones of the Fountain Formation are magnificently exposed (See Figures 88 and 89).

Along the east side of the park, river channel sandstones of the Permian Lyons Formation outcrop, followed by red *mudstones* of the Triassic/Permian Lykins Formation. Interbedded within the red, *nonmarine* mudstones are two thin, white *marine limestones* (See Figure 26). Farther to the east, the Dakota *Hogback* contains exposures of the Jurassic Morrison Formation, the Cretaceous Lytle Formation, the South Platte Formation and part of the Benton Formation. These rocks record a transition from nonmarine *floodplain* deposits to *marginal marine* to marine as the ocean began to advance over the land.

Dinosaur bones may be viewed in the Morrison Formation and dinosaur footprints viewed in the South Platte Formation. The South Platte exhibits many other geologic features, including *cross-bedding*, *ripplemarks* and *trace fossils*. The Benton, exposed on the east side of the Dakota Hogback, is composed of marine *shales* which contain fossils of fish scales, fish bones, ammonites, and occasionally sharks teeth. In the Dinosaur Ridge portion of the Dakota Hogback, fossil collecting is prohibited, as this is a designated National Historical Landmark. Additionally, dinosaur bones may be collected from public lands only by permit.

A non-profit organization known as THE FRIENDS OF DINOSAUR RIDGE conducts tours and provides information regarding the various geologic features which may be viewed at Dinosaur Ridge. The "FRIENDS" has a Visitor Center on the east side of the ridge near the junction of North Rooney Road and Alameda Parkway.

East of the Dakota Hogback, there are exposures of the Late Cretaceous including the marine Pierre Shale, the marginal marine Fox Hills Formation and the nonmarine Laramie Formation. Green Mountain is comprised of the Cretaceous/Tertiary Denver/Arapahoe Formation and the Tertiary Green Mountain Conglomerate. These rocks record the regression of the Cretaceous Sea and the beginning of the uplift of the modern-day Rocky Mountains.

Figure 88. Red Rocks Park. In the center of the photograph lies the famous amphitheater built by the Civilian Conservation Corps (CCC) in the 1930's. Large *flatirons* of the Pennsylvanian Fountain Formation flank the amphitheater to the right and to the left. Precambrian gneisses of the Idaho Springs Formation are visible on the slope of the mountain in the background.

Figure 89. Red Rocks. Red sandstones of the Fountain Formation in the area of Red Rocks Park are shown flanking the gneisses of Mountain Morrison. The white ridge at the bottom is the Lyons Formation. The Lyons in most areas is reddish, but here it is white.

A different look at the sand dunes of the Great Sand Dunes National Monument. They were covered by snow from a snow storm on May 1 of 1999. The dunes are arid and do not often receive enough snow to completely cover them.

MINING IN COLORADO

Today, Colorado is world famous for its beautiful mountains. In the Nineteenth Century and on into the early Twentieth Century, Colorado was famous for its mines. Most of the mines are located in a northeast-trending zone known as the *Colorado Mineral Belt* (See Figure 41). Notable exceptions to this are the following mining districts:

Cripple Creek:	Gold
Silver Cliff:	Silver
Western Colorado:	Uranium and vanadium

The beginning of the mining industry in Colorado occurred with the discovery of *placer* gold in 1858 around Denver. The original discovery was made by the "Georgia *Argonauts*" led by William Russell Green. Some members of this party had California gold experience. They discovered gold in the Platte River some three and one-half miles above the confluence with Cherry Creek and in Dry Creek about two miles from the Platte (Del Rio, 1960, p. 4).

A relatively small amount of gold was obtained. However, it was enough to greatly excite the prospectors and the famous "Pikes Peak or Bust" gold rush began. The gold placers in the Denver area had nothing to do with Pikes Peak. But the only geographic landmark in this part of Colorado people back east were familiar with was Pikes Peak, hence the gold rush was named after it.

From the Platte River and Dry Creek placers, prospectors panned their way up Clear Creek. Traces of gold were followed resulting in subsequent discoveries at Idaho Springs, Gregory Gulch and Blackhawk in 1859. From these areas, prospectors fanned out over the rest of Colorado searching for new gold placers. One by one, the rest of Colorado's gold mining districts were discovered. Many of these mining districts were initially discovered by prospectors *panning* and *sluicing* for gold in streams (See Figure 90). Later, the source (mineralized veins) was discovered and underground *lode* mining began.

The primary (initial) ore deposits contained sulfide *minerals*. These commonly included *pyrite*, *sphalerite*, *galena*, and *chalcopyrite*. The sphalerite and galena frequently contained gold and silver as impurities. Pyrite in some areas carried values of gold. As *weathering* processes removed overlying rocks, descending ground water decomposed pyrite, forming sulfuric acid. Sulfuric acid then attacked other sulfide minerals, decomposing them. Secondary "oxide" minerals then formed, creating a zone of oxide minerals above the remaining deeper sulfide ores. This is a common situation wherever sulfide ore deposits are found. Lode mines commonly first encounter the oxide zone and as mining deepens, enter the underlying sulfide zone. This was the situation at many of the Colorado districts. Oxide minerals such as *cerrusite*, *cerargyrite*, *argentite*, *chalcocite* and some *native gold* and silver were encountered first. Later, as mining progressed deeper, primary sulfide minerals were found.

Figure 90. Gold Mining. Placer mining for gold in Boren Creek, a tributary to La Plata River northwest of Durango. Photo by W.H. Jackson, 1875. USGS Photo Library.

Colorado contains many mining districts in which mining occurred during the later half of the Nineteenth Century and early Twentieth Century. It is not possible to describe all of them in this publication. Only those of greater historical significance are mentioned.

Colorado's fabulous era of gold mining largely ceased with the advent of World War II. During the war, only those mines producing metals needed for the war effort, such as copper, lead and zinc, were allowed to operate. As a consequence, a majority of the gold mines were shut down. After the war, many never reopened. Production figures for gold, silver and other metals given are obtained from Vanderwilt (1947). Unfortunately, records of metals mined and sold are rather incomplete and not always organized in the same manner. The production values in terms of dollars are not given because the strength of the dollar was much different in those days. In terms of today's gold prices, the total value would be much higher. Prior to 1933, the price of gold was $20 per ounce and after 1933, it increased to $35 per ounce. Today, it generally varies near $300 per ounce. In terms of current gold prices, a million dollars in gold in the Nineteenth Century would today be worth between $15 and $20 million today!

BLACKHAWK, CENTRAL CITY AND IDAHO SPRINGS: 1859

With the advent of the spring thaw in 1859, prospectors began to pan to the west of Denver up Clear Creek, which contained gold dust (and still does today). *Placer* gold was soon found at what is now Idaho Springs and also in streams west of the city of Boulder. In May of 1859, John H. Gregory, one of the "Georgia *Argonauts*", made the most significant discovery to date on north Clear Creek near Blackhawk, when he found a gold-bearing vein. Many *lode* discoveries followed and Central City was soon established.

The veins initially were found to contain free (native) gold in the oxide zone. As the underground mines deepened, they penetrated the sulfide zone containing sulfide *minerals*, such as *pyrite*, *sphalerite*, *galena*, *chalcopyrite*, *tennantite* and also gold *tellurides*. Gold and silver were principally associated with *chalcopyrite* and tennantite, although some gold was associated with pyrite. Some *pitchblende*, a uranium ore, was found in certain veins.

Total production for Gilpin County (excluding Idaho Springs whose records are included in those for Clear Creek County) from 1859 to 1945:

Placer: 48,000 oz. gold.
           6,467 oz. silver.

```
Lode:      4,206,000 oz. gold.
           2,680,000 oz. silver.
          26,674,000 lb. copper.
          38,984,000 lb. lead.
             868,000 lb. zinc.
```

GEORGETOWN, EMPIRE AND SILVER PLUME: 1859

As prospectors continued the westward move up Clear Creek from Idaho Springs, mineralized veins were found at Georgetown, Empire and Silver Plume.

Lodes in the Georgetown area contained veins similar to those of Idaho Springs, Blackhawk and Central City and yielded gold, silver, copper, lead and zinc. Empire was largely a gold-producing area and veins contained gold-bearing *pyrite* and *chalcopyrite*. Silver Plume, as the name suggests, was primarily known as a silver producing area. *Mineral* deposits yielded silver, lead and zinc.

Total production for Clear Creek County (including Idaho Springs):

```
Placer:      140,000 oz. gold.
                 400 oz. silver.

Lode:      1,238,000 oz. gold.
           7,933,000 oz. silver.
          14,043,000 lb. copper.
          69,728,000 lb. lead.
          35,301,000 lb. zinc.
```

BOULDER: 1859

Early in 1859, *placer* gold was discovered in the Boulder area. Subsequently, many mining districts were developed. Early production records do not separate placer from *lode* production, but the majority was probably lode mining. In 1900, the element tungsten, contained in the *mineral ferberite*, was discovered in Boulder County. Tungsten is a valuable alloy used in special cutting tools and drills. It was mined from 1900 through 1945 and received government price supports as it was a strategic element during war times. From 1900 to 1945, 24 thousand tons of tungsten concentrates were produced.

Total production (lode and placer combined):

```
1,000,000 oz. gold.
8,600,000 oz. silver.
1,700,000 lb. copper.
8,800,000 lb. lead.
   88,000 lb. zinc.
```

LEADVILLE: 1859

   Leadville was the second largest mining district in the state. Initially, gold was found in *placers* and then mining shifted to veins or lodes. Mineralized *limestones* of Mississippian, Devonian and Ordovician age were found to have rich ores containing silver, gold, lead, zinc and copper. Most ore came from replacement bodies in *dolomite*. *Hydrothermal* solutions from Tertiary *igneous* intrusions moved through fractures into the dolomites and replaced them with sulfide *minerals*. The primary oxide minerals were *cerrusite*, *cerargyrite*, *argentite*, *chalcocite* and some *native gold* and silver. Important sulfide *minerals* were *sphalerite*, *galena*, *argentite*, *tetrahedrite* and *chalcopyrite*.

Total production from 1859-1944:

Placer:	344,000 oz. gold.
Lode:	2,467,000 oz. gold.
	236,610,000 oz. silver.
	2,066,908,000 lb. lead.
	1,428,076,000 lb. zinc.
	102,874,000 lb. copper.

   In terms of today's prices, the silver would be worth more than one billion dollars and the gold about 840 million dollars. Truly, these were world class deposits!

BRECKENRIDGE: 1859

   Breckenridge was primarily an area of *placer* gold mining involving *panning*, *sluicing* and *dredging*, although substantial amounts of silver, lead and zinc were produced by underground *lode* mining. In 1880, narrow, rich veins containing crystallized gold were found. The *geology* in the local area is composed of sedimentary rocks including the Maroon, Morrison, Dakota, Benton, Niobrara and Pierre formations. These were *faulted* and intruded by Tertiary *igneous rocks* and sulfide veins were emplaced into the igneous and sedimentary rocks. *Weathering* of the ore bodies produced oxide veins containing gold.

Total production from 1859 to 1942:

Placer:	750,000 oz. gold.
Lode:	33,000 oz. gold.
	1,500,000 oz. silver.
	200,000 lb. copper.
	60,000,000 lb. lead.
	170,000,000 lb. zinc.

FAIRPLAY: 1859

The Fairplay district was a *placer* gold area. The valley gravels are glacial outwash derived from east slopes of the Mosquito Range. *Sluicing*, *hydraulic mining* and *dredging* were used to recover gold from these deposits. The largest area dredged was immediately south of the town of Fairplay, where immense piles of gravels may still be seen. The first dredge was constructed in 1922. A larger one operated from 1941 until 1952, when it became unprofitable to operate. The dredge, which was quite scenic, sat south of Fairplay for several decades. Unfortunately, it was eventually sold and taken to Columbia, South America where it began dredging for gold again.

Separate production records for the Fairplay district are not available. Statistics are available for Park County as a whole, but this includes a number of mining districts. Production for Park County is as follows:

Placer: 264,000 oz. gold.
            21,000 oz. silver.

Lode: 1,012,000 oz. gold.
         7,627,000 oz. silver.
         3,052,000 lb. copper.
        45,477,000 lb. lead.
         6,332,000 lb. zinc.

SAN JUAN REGION: 1870

Much of the mining in the San Juan Mountains occurred near the towns of **Silverton**, **Ouray**, **Telluride*** and **Creede** in and around the Silverton Caldera. This is a large area and there are a number of mining districts in it. Early prospectors mined gold *placers* but the focus soon shifted to *lode* mining. The majority of the production came from veins and replacement bodies in *volcanic* rocks. Ore *minerals* were the usual sulfides containing gold, silver, lead, zinc and copper.

Total production from 1870 to 1944 for all mining districts:

Lode: 7,630,000 oz. gold
       218,479,000 oz. silver
       999,544,000 lb. lead
       386,333,000 lb. zinc
           122,582, lb. copper

*No one is quite sure how the town of Telluride got its name. It would be logical to assume it was derived from telluride minerals which are present in some of the other mining districts of

187

Colorado. However, tellurides were not found in the mines of the Telluride region. Many locals believe it came from miners who, when referring to the rigors of working in the mines of the area, exclaimed "to hell you ride".

SILVER CLIFF AND WESTCLIFFE 1872

Silver Cliff and Westcliffe were *lode* mining areas and produced silver, gold, lead, copper and zinc from *minerals* in veins emplaced in Tertiary *volcanic rhyolite* and Precambrian *granite*.

Total production from 1872 to 1923:

Lode Mining: 100,000 oz. gold.
4,600,000 oz. silver.
580,000 lbs copper.
40,000,000 lbs lead.
1,000,000 lbs zinc.

ASPEN DISTRICT: 1880

Aspen was famous for its rich silver *lodes*. A small amount of gold and substantial amounts of copper, lead and zinc were also produced. Ore bodies were found in Mississippian Leadville *dolomites*.

Total production from 1880 to 1945:

Placer Mining: 28,000 oz. gold.

Lode Mining: 101,000,000 oz. silver.
1,100,000 lbs copper.
756,000,000 lbs lead.
22,000,000 lbs zinc.

CRIPPLE CREEK: 1891

The last major gold discovery in Colorado was Cripple Creek and it was the largest. It was discovered late because there was no *placer* gold and no easily recognizable *minerals* such as *sphalerite*, *galena*, or *chalcopyrite*. Gold was contained in a relatively rare mineral called *telluride*. Actually, telluride is a group of minerals which contain gold and silver. They were found in veins emplaced in rocks around the rim of a Tertiary-age basin which is surrounded by Precambrian *granite*, *gneiss*, and *schist*.

Until 1935, geologists believed that the basin was a very large, explosive *volcanic caldera*. After 1935, geologic field studies revealed a different picture (Koschman, 1972, p.207-8). Initially, a basin was created by subsidence and was bounded on all sides by steep *faults* . Subsidence was intermittent. Non-volcanic *conglomerate*, *arkose* and *mudstone* with thin layers of *shale* and *limestone* were deposited in the basin. These rocks were derived from *weathering* and *erosion* of surrounding Precambrian *crystalline* rocks. They were transported by water and deposited in the subsiding basin. Later, the basin was filled by *volcanic tuff* and *breccia*. These rocks, although composed of volcanic materials, exhibit sedimentary features which clearly show they were reworked by *erosion* and deposited mainly under water. The basin was apparently a shallow lake for much of the time. The source of the volcanic sediments deposited in the basin is not known. However, rocks of the basin are cut by dikes and intruded by *intrusive igneous rocks*. Clearly, there was extensive igneous activity in the area. Gold-bearing veins were then emplaced largely around the rim of this basin.

Total production from 1891-1944:

Lode Mining: 18,497,000 oz. gold.
2,084,535 oz. silver.

CLIMAX: 1918 to Present

The famous Climax mine is located about thirteen miles northeast of Leadville, on the continental divide at Fremont Pass (11,318 ft. el.). It is a huge, low-grade deposit of *molybdenite*, (an ore of *molybdenum*) in Tertiary *intrusives* and Precambrian *granites*, *schists* and *gneisses*. Tertiary intrusives are believed to have fractured the surrounding older rocks and molybdenite was emplaced in the fractures. According to Del Rio (1960, p.318), the mine originally contained ore reserves of one billion pounds of molybdenum. It is used as a steel alloy.

WESTERN COLORADO: 1945 to Recent

Uranium and vanadium were mined over a very large area in the *plateau* province of western Colorado. These ores were found in *sedimentary rocks* of Mesozoic age. Uranium occurred largely as the *mineral carnotite*, and the vanadium in a mineral called *vanadinite*.

The uranium industry in Colorado had its beginning in 1871. In this year, *pitchblende* (a mineral containing uranium) was discovered in a mine in the Central City district. Relatively small amounts of pitchblende were mined and sold. In 1898, carnotite, another uranium mineral, was identified in sedimentary rocks in Montrose County. During that same year, Marie and Pierre Curie discovered the element, radium. They also found all uranium ores contained small amounts of this element. Marie Curie used

189

radium from the Central City ores for her famous experiments which resulted in the discovery of radioactivity.

After World War II, a huge uranium mining industry developed in western Colorado. It provided a considerable portion of the uranium for the country's nuclear weapon and energy programs. The largest area of uranium mining occurred in what is known as the Uravan Mineral Belt southwest of Grand Junction in western Mesa, Montrose and San Miguel counties. Since 1947 and through 1980, 87 million pounds of uranium oxide were mined (Kent, et al, 1980, p. 219). The uranium mining industry has declined seriously in recent years.

Today, environmental problems are associated with mining dumps and mill tailings left over from uranium mining. In fact, a new industry has been created to reclaim these areas.

PETROLEUM AND NATURAL GAS

Oil was discovered in Colorado late in the Nineteenth Century and as of 1998, 1.7 billion barrels of crude oil and 7.6 billion cubic feet of natural gas have been produced.

Eastern Colorado: Many oil fields were discovered and produced from Cretaceous and Paleozoic rocks. A few of the fields produced some tens of millions of barrels of oil and a number of them produced a few million barrels. The rest produced less than a million barrels each.

Rangely Field: Located south of Dinosaur National Monument, Rangely is the largest oil field in Colorado. 650 million barrels of oil have been produced, largely from a Permian sandstone (Weber Formation).

Many other areas of Colorado contain oil fields, including North Park Basin, Paradox Basin, northwestern Colorado and other places where significant thicknesses of sedimentary rocks occur.

OIL SHALE

In Colorado, Utah and Wyoming, the Green River Formation contains thick deposits of *oil shale*. The "shale" is actually a shaley *limestone* bearing large amounts of organic matter called *kerogen*. When heated, the shale yields crude oil from the kerogen.

Organic matter accumulated in an extensive Tertiary lake named Goishute, which no longer exists. In Colorado, this lake occurred in what is now known as the Piceance Basin. Today, the paleo-Piceance Basin is the Roan and Grand Mesa *plateaus*. In the Piceance Basin, oil shale underlies about 1,400 square miles and ranges from 15 to 2,000 feet in thickness. About one trillion barrels of crude oil is present in oil shales which contain 15 or more gallons of oil per ton (Del Rio, 1960, p.451). The United States today consumes about 6 billion barrels of oil per year, much of which is imported. At the present rate of use, oil shale could provide the energy needs of the United States for approximately 150 years!

The bad news is this type of oil is very expensive to obtain. Currently, for oil shale to be profitable, oil would have to sell for $75 to $100 per barrel. The present world price for a barrel of crude oil is less than $20. In the future, as energy supplies of the world grow short, oil shale may become economic and a very valuable resource for the United States. There are environmental problems as well. The oil shale would presumably have to be strip mined and enormous dumps of waste rock would be created. Water and air pollution might also occur.

COAL

Colorado contains an estimated 80 billion tons or more of economic coal (Del Rio, 1960, p. 465). Slightly more than one percent has been mined. Clearly, Colorado has substantial reserves of coal. Within the state, coal is largely found in rocks of Late Cretaceous and Early Tertiary age.

Coal is formed in extensive marshes and swamps where plant materials accumulate and later lithify into coal. Today, there are only very small marshes and swamps scattered about the state. Therefore no new coal deposits of any consequence are being formed in Colorado at the present time. How did Colorado obtain such large quantities of coal? The climate of the earth was different during Cretaceous and early Tertiary times. It was warmer and based upon plant fossils found, borderline subtropical to temperate. There were no polar icecaps and dinosaurs (during the Cretaceous) roamed polar areas of the planet. Palm trees grew near present day Anchorage, Alaska! During this time, Colorado was near sea-level and some parts were low and swampy. In these swamps, plant materials accumulated and later changed into coal. In summary, Colorado obtained its substantial coal deposits at a time when the landscape was very different from that of today.

Coal has been mined over many portions of the state. The largest deposits are found in the following areas:

Walsenburg-Trinidad Region: Substantial underground coal mining occurred in this area in the late Nineteenth and early Twentieth centuries.

Denver Region: This large area extends from the Wyoming State line southward to nearly Colorado Springs. It was heavily mined in the late Nineteenth and early Twentieth centuries. In some of these mines, underground coal fires were a serious problem and today, ground subsidence in urban areas located over old coal mines is a threat.

Northwestern Colorado: Large reserves exist, some of which are currently being mined for electric power generation.

THE END

A large drape fold (monocline) along the southern side of Dinosaur National Monument. Note the rock layers are horizontal on the left, but as they approach the right side, they dive steeply, forming the monoclinal structure. This may be viewed from U.S. Highway 40 between Vernal, Utah and Dinosaur, Colorado.

SELECTED REFERENCES

American Assoc. Petroleum Geologists, 1967, Geological Highway Map, Southern Rocky Mountain Region, Utah-Colorado-Arizona-New Mexico: American Assoc. Petroleum Geologists, Tulsa, OK.

Barrs, D.L., and J.A. Ellington, 1984, Geology of the Western San Juan Mountains, in D.C. Brew, ed., Field Trip Guidebook: Geological Society of America, Rocky Mountain Section, 37th Annual Meeting, p.1-45.

Blair, Rob, Managing Editor, and Casey, T.A., Romme, W.H., and Ellis, R.N., Technical Editors, 1996, The Western San Juan Mountains: University Press of Colorado.

Bloom, A.L., 1998, Geomorphology, 3rd. ed.: Prentice Hall, Upper Saddle River, N.J., p.129-130.

Chronic, J., 1976, Diamond-bearing Paleozoic Diatremes in Colorado and Wyoming, in Professional Contributions of Colorado School of Mines: No. 8, p.101-9.

Chronic, John, McCallum, M.E., Ferris, C.S., and Eggler, D.H., 1969, Lower Paleozoic Rocks in Diatremes, Southern Wyoming and Northern Colorado, in Geological Society of America Bulletin: Vol. 80, p.149-156.

Cooper, J.D., Miller, R.H., and Patterson, J., 1990, A Trip Through Time: Merrill Publishing Company, Columbus, Ohio.

Del Rio, S.M., 1960, Mineral Resources of Colorado, First Sequel: Mineral Resources Board, State of Colorado, Denver, Colorado.

Epis, R.C., Scott, G.R., Taylor, R.B. and Chapin, C.E., 1980, in Kent, H.C. and Porter, K.W., editors, Colorado Geology: 1980 Symposium, Rocky Mountain Association of Geologists, Denver, CO.

Griffitts, M.L., 1990, Guidebook to the Geology of Mesa Verde National Park, Colorado: Mesa Verde Museum Association, Mesa Verde National Park.

Grose, L.T., 1972, Geologic Formations and Structure of Colorado Springs Area, Colorado in The Mountain Geologist, V.9, No. 2-3. The Rocky Mtn. Assoc. of Geologists, Denver, Colorado.

Haber, Heinz, 1969, Our Blue Planet; the Story of the Earth's Evolution: Scribner, N.Y.

Hansen, W.R., 1977, Geologic Map of the Canyon of Lodore South Quadrangle, Moffat County, Colorado: Map GQ-1403. U.S. Geol. Survey.

Hansen, W.R., 1977, Geologic Map of the Jones Hole Quadrangle, Uinta County, Utah, and Moffat County, Colorado: Map GQ-1401. U.S. Geol. Survey.

Hansen, W.R., Carrara, P.E., and Rowley, P.D., 1982, Geologic map of the Canyon of Lodore North Quadrangle, Moffat County, Colorado: Map GQ-1568. U.S. Geol. Survey.

Hausel, W.D., McCallum, M.E., and Woodzick, T.L., 1979, Exploration for Diamond-Bearing Kimberlite in Colorado and Wyoming: An Evaluation of Exploration Techniques: Report of Investigations No. 19, The Geological Survey of Wyoming.

Hildebrand, A.R., 1993, The Cretaceous/Tertiary Boundary Impact (or the Dinosaurs Didn't Have a Chance): Journal Royal Astronomy Society of Canada, v.87, no.2, p. 77-117.

Howard, J.D., 1966, Patterns of Sediment Dispersal in the Fountain Formation of Colorado in The Mountain Geologist, v.3, no.4, p.147-153, Rocky Mtn. Assoc. Geologists, Denver, CO.

Hunt, C.B., 1969, Geologic History of the Colorado River: Prof. Paper 669, U.S. Geol. Survey.

Johnson, R.B., 1968, The Great Sand Dunes of Southern Colorado: in The Mountain Geologist, v.5, no.1, p.23-9, Rocky Mtn. Assoc. of Geologists, Denver, CO.

Jenkins, J.T.,Jr., and Jenkins, J.L., 1993, Colorado's Dinosaurs: Special Publication 35, Colorado Geological Survey, Denver, CO.

Kaye, Glen, 1987, Chief Park Naturalist, Rocky Mountain National Park, on reverse side of Rocky Mountain National Park 1:50,000-Scale Topographic Map, National Park Series, U.S. Geol. Survey.

Kirkely, M.B., Gurney, J.J., and Levinson, A.A., 1991, Age, Origin, and Emplacement of Diamonds: Scientific Advancements in the Last Decade, in Gems and Gemology, Journal of the Gemological Institute of America, Vol. 27. No. 1, p.2-25.

Koschman, A.H., 1972, Cripple Creek District, in The Mountain Geologist, v. 9, nos. 2-3, p. 207-209, Rocky Mtn. Assoc. Geologists, Denver, CO.

Kyte, Frank, 1996, The Rock that Killed the Dinosaurs?: in Earth, August, 1996, p. 12-13.

LeRoy, L.A., 1960, Generalized Composite Stratigraphic Sections of Western Colorado: Colorado School of Mines Department of Publications, Golden, CO.

Lipman, Peter, 1997, Chasing the Volcano: in Earth, December, 1997, p.32-39.

Nichols, D.J. and Fleming, R.F., 1990, Plant Microfossil Record of the Terminal Cretaceous Event in the Western United States and Canada, in Special Paper 247, Geological Society of America, p.445-455.

MacGinite, H.D., 1953, Fossil Plants of the Florissant Beds, Colorado: Carnegie Institution of Washington Publication 599, Washington, D.C.

Mallory, W.W., 1958, Pennsylvanian Coarse Arkosic Redbeds and Associated Mountains in Colorado: in Symposium on Pennsylvanian Rocks of Colorado and Adjacent Areas, Rocky Mtn. Assoc. Geologists, Denver, CO.

Mallory, W.W., 1972, Pennsylvanian Arkose and the Ancestral Rocky Mountains: in Geologic Atlas of the Rocky Mountain Region, Rocky Mtn. Assoc. Geologists, Denver, CO.

Meyer, H.W., Weber, Laine, 1995, Florissant Fossil Beds National Monument: Preservation of an Ancient Ecosystem, Rocks and Minerals, Vol.70, No. 4, p.232-9.

Miller, W.E., Baer, J.L., Stadtman, K.L. and Britt, B.R., 1991, The Dry Mesa Quarry, Mesa County, Colorado: in Guidebook for Dinosaur Quarries and Tracksites Tour: Western Colorado and Eastern Utah, 1991, Grand Junction Geological Society, P. 31-46.

Pillmore, C. L., Fleming, R.F. and Nichols, D.J., 1994, Field Guide to the Continental Cretaceous-Tertiary Boundary in the Raton Basin, Colorado and New Mexico (unpublished): Field Trip # 10, American Assoc. Petrol. Geol., Annual Meeting, Denver, CO.

Preston, R.J., 1961, North American Trees, Revised Ed.: M.I.T. Press.

Recer, Paul, 1997, Seabed Yields "Smoking Gun" in Asteroid Hit: in Rocky Mountain News, February 17, 1997. Denver, CO.

Rigby, J.K., 1977, Southern Colorado Plateau: K/H Geology Field Guide Series. Kendall/Hunt Publishing Company.

Rold, J.W., 1977, The Marble Area - A Development Frontier 1873-1977, in Veal, H.K., ed., 1977 Symposium, Exploration Frontiers of the Central and Southern Rockies. Rocky Mtn. Assoc. of Geologists, Denver, CO.

Sanders, G.F., Jr., Scott, G.R., and Naeser, C.W., 1976, The Buffalo Peaks Andesite of Central Colorado. U.S. Geological Survey Bulletin 1405-F.

Soister, P.E., 1978, Stratigraphy of Uppermost Cretaceous and Lower Tertiary Rocks of the Denver Basin, p.223-230, in 1978 Symposium, Rocky Mtn. Assoc. Geologists, Denver, CO.

Steven, T.A., and P.W. Lipman, 1976, Calderas of the San Juan Volcanic Field, Southwestern Colorado: Professional Paper 958. U.S. Geological Survey.

Thompson, S.E., 1996, "Dazed and Duped by Diamonds", in Lapidary Journal, Vol. 50, No. 4, p.59-66, 108.

Tweto, Ogden, 1979, Geologic Map of the State of Colorado: U.S. Geol. Survey.

Tweto, Ogden, 1980a, Tectonic History of Colorado: in Kent, H.C. and Porter, K.W., editors, Colorado Geology, in 1980 Symposium, Rocky Mtn. Assoc. Geologists, Denver, CO.

Tweto, Ogden, 1980b, Summary of Laramide Orogeny in Colorado: in Kent, H.C. and Porter, K.W., editors, Colorado Geology: 1980 Symposium, Rocky Mtn. Assoc. Geologists, Denver, CO.

U.S. Department of Energy, 1980, Carbon Dioxide Effects Research and Assessment Program, Washington, Carbon Dioxide and Climate Division, Report 008.

Vanderwilt, J.W., 1947, Mineral Resources of Colorado: Mineral Resources Board, State of Colorado, Denver, Colorado.

Voynick, S.M., 1994, The Moffat County Diamond Fields, p.243-7: in Colorado Rockhounding, Mountain Press Publishing Company, Missoula, Montana.

Walker, T.R., and J.C. Harms, 1972, *Eolian* Origin of Flagstone Beds, Lyons Sandstone (Permian), Type Area, Boulder County, Colorado: in The Mountain Geologist, v. 9, nos. 2-3, p. 279-288, Rocky Mtn. Assoc. Geologists, Denver, CO.

Witze, Alexandra, 1996, The Rock That Killed the Dinosaurs: Earth, August, 1996, p.12.

GLOSSARY

accretion (1) The addition of sediments to a site of deposition. (2) The growth of continents by scraping sediments from a descending plate at a *subduction zone*, where a continental plate collides with an oceanic plate.

alcove Large, deep niche formed in a precipitous face of rock.

alluvial fan Fan-like accumulation of sediment deposited where a stream slows down at a relatively flat valley floor or when it enters a lake or ocean.

alluvial Pertaining to or composed of *alluvium*, or deposited by a stream or running water.

alluvium General term for clay, silt, sand or gravel deposited during comparatively recent geologic time by a stream or other body of running water as a sorted or semisorted sediment in the bed of a stream or on its *floodplain* or *delta*.

alpine glacier Any glacier in a mountain range except an ice cap or an ice sheet. It usually originates in a *cirque* and may flow down into a valley previously carved by a stream.

ammonia Chemical compound containing nitrogen and hydrogen (NH_3).

ammonite Shell-bearing organism, usually coiled but sometimes straight-shelled. Similar to the modern-day Nautilus. Ammonites are members of the Cephalopod Class of the Mollusk Phylum. Many thousands of species are known. Sizes range from very tiny to many feet in diameter. Reached greatest diversity in the Jurassic and Cretaceous periods. Became extinct at the end of the Cretaceous.

amphibole Dark colored minerals which are members of the amphibole group. These are chain silicates and their chemistry is highly variable. *Hornblende* is the most common. Amphiboles are mostly found in intermediate composition igneous rocks and in high temperature metamorphic rocks.

Anasazi The name given to the peoples inhabiting the cliff dwellings in the general vicinity of Mesa Verde National Park. The name has been said to mean the "Ancient Ones". Another definition of the name is "The Enemy".

andesite Fine-grained *volcanic rock intermediate* in composition between *rhyolite* and *basalt*, consisting of about equal amounts of *feldspar* and *ferromagnesian minerals*.

andesitic An adjective, generally used to describe rocks that are largely composed of andesite.

anticline Convex upward *fold* in rocks.

aragonite Calcium carbonate ($CaCO_3$). Aragonite has the same chemical composition as *calcite* but crystallizes in the orthorhombic crystal system while calcite is in the hexagonal system. Aragonite is an unstable mineral and slowly converts (in its crystal structure) to calcite.

arete Narrow, jagged, serrate mountain crest, or a narrow, rocky, sharp-edged ridge or spur. Commonly present above snowline in rugged mountains sculptured by glaciers. Resulting from the continued backward growth of the walls of adjoining *cirques*.

argentite Silver sulfide mineral (Ag$_2$S), and an important ore of silver.

argonaut Gold seeker.

arkose *Feldspar*-rich sandstone.

arkosic Said of a sandstone containing minor amounts of *feldspar*. For example, an arkosic sandstone.

ash *Extrusive igneous* unconsolidated material or *rock* composed of fine *pyroclastic* fragments under 2 millimeter in diameter.

ash flow Turbulent blend of unsorted, mostly fine-grained *pyroclastics* and high-temperature gas ejected explosively from fissures or a crater.

asthenosphere The portion of the upper *mantle* beneath the *lithosphere*. It consists of hot, plastic rock where *magma* may form. Extends from a depth of about 100 kilometers to about 350 kilometers below the earth's surface.

augite The most common mineral of the *pyroxene* mineral group, which are chain silicates of complex chemistry. Augite is usually black, greenish black or dark green and is abundant in *mafic igneous rocks* and in certain *metamorphic rocks*.

Bacculites Straight-shelled ammonite.

bajada Broad, continuous alluvial slope or gently inclined detrital surface extending along and from the base of a mountain range into and around an inland basin. Formed by the lateral coalescence of a series of separate alluvial fans. It occurs most commonly in arid or semiarid regions like the southwestern United States.

barrier island Long, low, narrow, wave-built sandy island parallel to but separated from the mainland by a *lagoon*. Padre Island and Galveston Island along the coast of Texas are good examples of barrier islands.

basalt Dark-colored, very fine-grained *volcanic rock* composed of *feldspar* and heavy *ferromagnesian minerals*. Derived from *magma* in the asthenosphere.

basement rock General term used for the rocks underlying the oldest identifiable rocks in the area, or rocks of the earth's *crust* below sedimentary deposits, extending downward to the top of the mantle. In many places, these rocks are *igneous* and *metamorphic* and are Precambrian in age.

batholith Large *intrusive igneous* body with an outcrop area greater than 100 square kilometers.

beach Strip of shoreline washed by waves or tides. Frequently composed of sand.

bioclastic *Limestones* and *dolomites* composed of fragments of shells and hard parts of organisms.

biotite Black, platy *mineral* of the *mica* group.

brachiopod Any solitary *marine* invertebrate belonging to the phylum Brachiopoda. A *shellfish* similar to a *clam* but with different shell structure. Most abundant during the Paleozoic Era. Almost became extinct at the end of the Permian Period and only one species survives today.

braided stream A stream that divides into a tangled network of several, small, branching and reuniting channels separated from each other by islands or channel bars. Resembling the strands of a complex braid.

breccia Coarse-grained *clastic* rock composed of large, angular and broken rock fragments cemented together in a finer-grained matrix.

burrow Hole dug into sediment by feeding organisms and sometimes used as living space. As an example, modern-day crabs dig holes in which they hide. Lithified rocks sometimes exhibit burrows, the presence of which may be used by geologists to interpret *depositional environments*.

B.Y. Billion years.

B.Y.A. Billion years ago.

calcareous Containing calcium carbonate ($CaCO_3$).

calcite *Mineral* composed of calcium carbonate ($CaCO_3$). *Limestone rocks* are composed of calcite grains and cement.

caldera Large, basin-shaped *volcanic* depression, more or less circular in form, the diameter of which is many times greater than the included vent or vents through which eruptions take place. Calderas generally have diameters of many miles and sometimes tens of miles. They are formed by the largest and most violent volcanic eruptions.

carnivore Meat-eating organism.

carnotite Secondary radioactive *mineral* containing uranium.

cephalopod An invertebrate with an external or internal shell, or sometimes no shell. Those with external shells may have a straight, slightly curving or coiled shell, all with internal chambers. Ammonites and bacculites are the most common cephalopods.

cerargyrite Silver chloride (AgCl) and an ore of silver.

cerrusite Colorless, white, yellowish or grayish *mineral* of the aragonite group ($PbCO_3$). It is a common alteration product of *galena* in oxidation zones of mineral deposits and is an important ore of lead.

CH_4 The chemical composition of *methane*, a compound of carbon and hydrogen.

chalcocite Copper sulfide mineral (Cu_2S) and a valuable ore of copper.

chalcopyrite Bright, brass-yellow *mineral* ($CuFeS_2$). It is generally found massive (not exhibiting crystal faces) and constitutes the most important ore of copper.

chemical rocks Rocks formed by the precipitation of *minerals* from water.

chert A variety of quartz. Hard, dense microcrystalline *rock* composed of microscopic interlocking crystals of quartz. Generally forms in *marine* sedimentary environments.

chrome diopside Bright-green variety of *diopside* containing a small amount of chromium. It is frequently present in *kimberlites* and is regarded as an indicator *mineral* of such.

cinder cone Conical volcanic hill formed by the accumulation of cinders or particles of *scoria*, normally of basaltic or andesitic composition.

cirque Deep, steep-walled, flat or gently floored half-bowl-like recess or hollow. Variously described as horseshoe or cresent-shaped or semicircular in plan. Situated high on the side of a mountain and commonly at the head of a glacial valley. Produced by the erosive activity of mountain glaciers. It often contains a small, round lake and it may or may not be occupied by ice or snow.

clam Informal name which may be applied to all *pelecypods*. Many fresh-water and a few marine clams are called mussels. Scallops and oysters are special types of clams. Clams evolved in the Paleozoic and became very abundant in the Mesozoic and Cenozoic eras.

clastic Composed of clasts. A clast is a fragment of a previously existing *mineral* or rock which has been transported some distance from its original source. The most common clastics are sandstones and shales.

clay (1) *Clastic mineral* particle of any composition having a diameter less than 1/256 millimeter. (2) *Clay* mineral.

claystone Hard, compacted *clay* rock having the texture and composition of *shale* but lacking its fine lamination or fissility.

cliff-former Said of horizontal rock layers that are resistant to *weathering* and *erosion* and which tend to form cliffs adjacent to valleys or plains. Sandstones, *limestones*, *dolomites* and basalt flows are common cliff-formers.

Colorado Mineral Belt Linear geographic area extending from the general vicinity of Silverton to Boulder which contains a majority of the metal mines of Colorado. This area was mineralized by *hydrothermal* solutions during *volcanic* episodes of the Tertiary.

comet Relatively small celestial object composed of rock and ice which has a "tail" composed of water vapor and/or dust.

concretion Hard, compact aggregate of *mineral* matter, subspherical to irregular in shape, formed by precipitation from water solution around a nucleus, such as shell bone or organic matter, in *sedimentary* or *pyroclastic rocks*. Common minerals involved include silicon and iron minerals such as chert, *hematite*, limonite and *pyrite*.

conglomerate Coarse-grained *clastic sedimentary rock* composed variously of pebbles, cobbles and boulders with a matrix of silt and sand.

continental drift The theory which proposed North and South America were once joined together with Europe and Africa in a large supercontinent named *Pangaea*. The continents subsequently "drifted" apart to their present positions. Originally postulated by the German meteorologist Alfred Wegener in 1912. Today, part of his work is incorporated in the *Plate Tectonics Theory*.

continental glacier Glacier of considerable thickness completely covering a large part of a continent, such as the ice sheets covering Greenland and Antarctica.

convection cell Moving, circulating matter carrying heat from an area which is hotter to an area which is cooler, where the heat is dispelled. A pot of boiling water contains convection cells.

coral   General name for any of a large group of bottom-dwelling *marine* invertebrate organisms attached to a base and belonging to the class Anthozoa. Corals are common in shallow areas of warm intertropical modern seas and are abundant in the *fossil* record in all periods later than the Cambrian. They produce external skeletons of calcium carbonate and can exist as solitary individuals or grow in colonies.

Cretaceous Seaway   Large, shallow sea that covered the interior of the North American continent during much of the Late Cretaceous Period. It extended from the Arctic Ocean to the Gulf of Mexico and divided the continent into two smaller continents.

crevasse   Deep and nearly vertical split, fissure, crack or rift in a glacier. Caused by stresses resulting from differential movement over an uneven surface.

cross-bedding   Cross-stratification or cross-lamination of the interior of a sand dune. Sand dunes are composed of very thin layers (lamina) of sand deposited at a steep angle relative to horizontal. Rocks which contain cross-bedding are said to be cross-bedded.

crust   The outermost layer of the earth. Continental crust is largely composed of silica-rich rocks such as *granite*. Oceanic crust is composed of basalt.

crystalline   Said of a rock consisting wholly of crystals or fragments of crystals. Especially *igneous rock* developed by cooling from a molten state and containing no glass. Also *metamorphic rock* that has undergone recrystalization as a result of temperature and pressure changes.

cuesta   Hill or ridge with a gentle slope on one side and a steep slope on the other.

deflation   Form of *erosion* involving the sorting out, lifting and removal of loose, dry, fine-grained particles (*clay* and silt sizes) by wind. Common in desert areas.

delta   Projection in a coastline created by sediment (sand, silt and *clay*) supplied by a river. *Deltas* commonly contain many subenvironments such as distributary river and tidal channels, bays, tidal flats, marshes, *swamps*, lakes and *lagoons*.

depositional environment   Any setting in which sediment is deposited. For example, sediment is deposited in rivers, lakes, marshes, swamps, beaches and many other settings. Sediments deposited in different depositional environments will have unique and different characteristics.

diatreme   Breccia-filled volcanic pipe formed by a gaseous explosion.

dike   Tabular *igneous* intrusion which cuts across the planar structures of the surrounding *rock*.

diopside   *Mineral* of the *pyroxene* mineral group [$CaMg(SiO_3)_2$].   Varies in color from white to green. Diopside occurs in some *metamorphic rocks* and especially as a contact-metamorphic mineral in *crystalline limestones*.

dim dwarf star   Class of stars which are small in size, high in gravity and emit only small amounts of light.

diorite   Coarse-grained *igneous intrusive rock* of *intermediate* composition between *granite* and *gabbro*.

dolomite A *mineral* and also commonly a *rock*. The mineral dolomite is a calcium magnesium carbonate [CaMg(CO$_3$)$_2$]. It may be formed (crystallized) in *sedimentary*, *metamorphic* and *igneous* environments. In sedimentary environments, dolomite is a common *marine* rock.

dredging Method of mining *placer* gold. The dredge is a large, floating machine with buckets and conveyor belts. It scoops up gravels from a placer deposit and runs it through an internal sluice, separating gold, and discharging the gravel behind it.

earthquake Sudden motion, movement, or trembling of the earth. Usually caused by shifting of rocks along a *fault*.

echinoderm An organism belonging to the phylum Echinodermata, which includes starfishes, sand dollars, sea urchins, sea cucumbers and crinoids.

elemental carbon Carbon which is not combined with other elements. Created by combustion. Smoke and soot from fires contain elemental carbon.

emergent Above sea level.

eolian Pertaining to wind. Wind erodes, transports and deposits fine-grained *clastic* sediments such as *clay*, silt and sand. Sand dunes are a common product of eolian processes.

erratic Rock fragment carried by a glacier and deposited as the ice melts. Usually occurring some distance from the outcrop where the fragment was derived. Generally of boulder size.

erosion The wearing away of soil and rock by *weathering* processes.

evaporite Nonclastic *sedimentary rock* composed primarily of *minerals* precipitated from a saline solution concentrated by evaporation. Especially a deposit of minerals precipitated from a *restricted* or enclosed body of seawater or from water of a salt lake. Examples include *gypsum*, anhydrite and rock salt.

extinction The total disappearance of an organism so that it no longer exists.

extrusive Class of *igneous rocks* which have been extruded onto the surface of the earth from a *volcanic* vent. The term extrusive is used to distinguish igneous rocks formed in this manner from those which are *intrusive*.

fault Fracture in rocks along which movement of rocks on one side of the fault relative to rocks on the other side has occurred.

fault block mountain Mountain which is an uplifted *fault* block.

fault block Block of rock bounded by *faults*.

feldspar Group of important *rock* forming *minerals* in the silicate mineral group. Silicates contain silicon oxides in their chemical composition. Feldspars may be subdivided into two important subgroups: those containing potassium (orthoclase and microcline) and those containing calcium and sodium (plagioclase series).

felsic Said of an *igneous rock* composed largely of light-colored *minerals*, such as *quartz* and *feldspar*. It is the opposite of *mafic*. *Granite* and *rhyolite* are felsic rocks.

ferromagnesians *Minerals* containing iron and magnesium such as *amphiboles*, *pyroxenes*, *biotite* and *olivine*. Rocks in the *mantle* are composed of ferromagnesians, typically olivine and *augite*.

felsenmeer Flat or gently sloping area covered with a continuous veneer of large angular blocks of rock derived from well-jointed underlying bedrock. Joints develop by intensive frost action. This process usually occurs on high, flat-topped mountains or *plateaus* above timberline.

ferberite Member of the wolframite group of *minerals*, which form a series from ferberite ($FeWO_4$) to hubnerite ($MnWO_4$). An ore of tungsten.

flatiron Triangular or wedge-shaped block of resistant rock which is steeply inclined from horizontal. Formed by *weathering* and *erosion* of steeply inclined strata, frequently flanking uplifted areas such as mountains.

floodplain Any flat or nearly flat area bordering a stream which may be covered by its waters at flood stages.

flow rocks Term applied to *volcanic rocks* which originated as *lava flows* from *volcanoes* or volcanic vents.

fold Curve or bend of layered rocks. A fold is usually the product of compressive deformation.

formation Mappable unit of rocks. More specifically, a contiguous group of rocks which exhibit the same readily identifiable characteristics.

fossil pollen Pollen from flowering plants is commonly deposited in fine-grained sediment such as *clay*. Consequently, fossil pollen is present in *shales*. It is very useful for age dating late Mesozoic and Cenozoic rocks. It also yields important paleo-climatic information.

fossil Evidence in the rocks of former life. The most abundant fossils are those of shells and skeletal parts of animals. Shells and skeletal parts may be partly or wholly replaced by *minerals* or may consist entirely of the original shell or bone material. More uncommonly, soft parts of plants and animals may be preserved. *Trace fossils* are a special class of fossils such as tracks, trails, and other disturbances made in sediment by living organisms.

fossiliferous Containing fossils.

gabbro Group of dark-colored, basic *intrusive igneous rocks* composed principally of calcium-rich feldspar and *augite*. Compositionally, gabbro is the approximate intrusive equivalent of *extrusive* basalt.

galena *Mineral* containing lead and sulfur (PbS). It commonly contains silver as an impurity. Galena is the most important ore of lead and frequently an important ore of silver.

garnet Group of silicate *minerals* which have varying amounts of calcium, magnesium, manganese, iron, aluminum and chromium. Specific garnets are almandine (red), andradite (green-brown), grossularite (yellow), pyrope (red), spessartine (red-orange) and uvarovite (green). Garnet occurs in *igneous* and *metamorphic rocks*.

gastroliths Polished stone or pebble from the stomach of some vertebrates and thought to have been used to grind their food. Sometimes informally referred to as "gizzard stones".

gastropod Any *mollusk* belonging to the Class Gastropoda: e.g. a snail.

geology The study of the planet earth. Geology is concerned with the origin of the planet, the material and morphology of the earth, and its history and the processes that acting upon it, yielding our present earth. Geology today also encompasses study of other planets of the solar system.

glacial interlude Brief span of time (usually about 10,000 years) between the major glacial periods of the Pleistocene Glaciation. During the interludes, glaciers retreat and the average temperature of the earth rises.

glaciation The formation, movement and recession of glaciers or ice sheets.

glauconite Dull-green, amorphous and earthy or granular *mineral* of the *mica* group. Occurs in sandstones of *marine* origin. It is believed to be derived from the chemical constituents of fish excrement contained in the original sediments. The presence of glauconite in sandstones is widely regarded as evidence of marine origin.

gneiss Foliated *metamorphic rock* which contains alternating layers of different *minerals*. The most common type of gneiss is quartz- and feldspar-rich, along with lesser amounts of iron-rich minerals such as *biotite* and *hornblende*. Gneisses frequently are intricately folded.

gradient Term used to describe the degree of inclination of a surface, such as a stream bed. It is expressed as a ratio of vertical change over some horizontal distance.

granite Coarse-grained *intrusive igneous rock* composed of *quartz* and *feldspar* with lesser amounts of iron-rich *minerals* such as *biotite* and *hornblende*.

granitic Pertaining to or composed of granite.

gypsum *Mineral* containing calcite sulphate plus water ($CaSO_4X2H_2O$). Gypsum is most commonly associated with *marine* evaporites (chemical *rocks*) such as halite and anhydrite where it may occur as thick layers. In large layers, it is a *rock*.

gypsiferous Containing gypsum.

hanging valley Glacial valley whose mouth is at a relatively high level on the steep side of a larger glacial valley. The discordance in level of their floors is due to the greater erosive power of the glacier in the larger glacial valley.

heat convection cells Heat convection is the transfer of heat by moving matter from a place where it is hot to a place where it is cool, dissipating the heat. A pot of boiling water contains heat convection cells.

hematite An iron oxide *mineral* (Fe_2O_3). Hematite may occur in *igneous*, *metamorphic* and *sedimentary* environments. Iron is a very active and highly mobile element easily dissolved in water. As such, it is a common precipitate mineral or cement in sedimentary rocks.

herbivore Plant eating organism.

high-angle fault A *fault*, the dip (inclination from horizontal) of which is greater than 45^0.

high-velocity shock fracture Rare type of fracture in the *mineral quartz*. Quartz normally has a conchoidal or irregular fracture. When it is subjected to high-velocity impact or shock, such as the explosion of an atomic bomb or by meteorite impact, fractures are created which consist of multiple, parallel planar surfaces which may overlap in a herringbone pattern. The presence of high-velocity impact fractures in quartz (referred to as high-velocity shocked quartz) is considered by many scientists to be evidence of high-velocity meteorite impact.

hogback Any ridge with a sharp summit and steep slopes of nearly equal inclination on both flanks. Resembles in outline the back of a hog. Specifically a long narrow, sharp-crested ridge formed by the outcropping edges of very steeply inclined or highly tilted resistant rocks. Produced by differential *erosion* of soft versus hard rocks.

horn High, rocky, sharp-pointed, steep-sided, pyramidal mountain peak with prominent faces and ridges. Bounded by intersecting walls of three or more *cirques* carved into a mountain by headward *erosion* of glaciers.

hornblende Most common mineral of the *amphibole* group of chain silicate minerals. Commonly black and a constituent of felsic to intermediate igneous rocks and metamorphic gneiss and schist. Uncommon in mafic rocks.

hornfels Fine-grained dark *rock* which generally will scratch glass. May have a few coarser *minerals* present. Commonly formed by contact *metamorphism* of *shale* or *basalt* when intruded by a body of *igneous magma*.

hydraulic mining Use of strong jets of water in *placer* mining. Gravels are washed down and pass through a sluice where gold is extracted.

hydrothermal Of or pertaining to heated water, the action of heated water, or the products of such action. For example, a *mineral* deposit precipitated from a hot aqueous solution.

igneous pipe Vertical conduit in the earth's *crust*, through which magmatic materials have passed. It is usually filled with *volcanic* breccia and fragments of older rock.

igneous *Rocks* or *minerals* which have solidified from molten *magma* are said to be igneous. The term is derived from the Latin word ignis, meaning "fire".

ilmenite An iron-black, opaque *mineral* ($FeTiO_3$). It is the principal ore of titanium. Occurs as a common accessory mineral in basic *igneous rocks* such as *gabbros*.

impermeable Said of an earth material (such as sediments or rocks) lacking the ability to transmit a fluid through internal voids or pore spaces.

Inoceramus Large *marine fossil clam*. Inoceramids were abundant in the Jurassic and Cretaceous periods. Some are very useful as index fossils for relative age-dating and rock correlations.

intermediate (rock) Said of an *igneous rock* whose composition is between *felsic* and *mafic*, such as *andesite* or *diorite*.

intermontane Situated between or surrounded by mountains, mountain ranges or mountainous regions.

intrusive Coarse-grained *igneous rock* which cooled and crystallized into hard rock <u>beneath</u> the surface of the earth.

inversion of relief A phenomenon in which differential *erosion* acts to change the relief of an area. Existing high areas may be eroded down and adjacent low areas more resistant to erosion then become the high ones. In *fold* belts, rocks in anticlines contain more fractures than do those in synclines and the anticlines consequently are less resistant to erosion than the synclines. The anticlines erode down, becoming valleys. The more resistant synclines become ridges.

iridium Rare heavy element which is a member of the platinum group of metals. Iridium is very scarce in rocks of the earth's *crust*.

island arc Curved chain of *volcanic* islands rising from the sea floor. They are near to continents and landward of *oceanic trenches* at *subduction zones*.

K-T Boundary Time boundary between older Cretaceous and younger Tertiary rocks. The letter K is used by geologists as a symbol or abbreviation for Cretaceous. The letter T stands for Tertiary.

kerogen Fossilized, insoluble, organic material found in *sedimentary rocks*, usually *shales* or shaley *limestones*, which can be converted by distillation to petroleum products.

kimberlite Peridotite containing abundant *olivine*, commonly decomposed to *serpentine*. Frequently contains *mineral* grains of *phlogopite*, chrome *diopside*, calcite, red pyrope *garnet*, *ilmenite* and *spinel*. Kimberlite is found in *diatremes* (volcanic pipes) which have their roots deep in the *mantle*. Kimberlites sometimes, but not always, contain diamonds.

lagoon Shallow body of water partly or completely separated from the sea by a reef or barrier island.

lahar Mudflow composed chiefly of *volcaniclastic* materials on the flank of a *volcano* or in areas where *volcanic ash* has been deposited.

landform Any physical, recognizable form or feature of the earth's surface, having a characteristic shape and produced by natural causes. Includes major forms such as a plain, *plateau* or mountain, and minor forms such as a hill, valley, esker or dune.

landslide General term covering a wide variety of mass movement *landforms* and processes involving moderately rapid to rapid transport of earth materials downslope by means of gravity. Landsliding includes such things as rockslides, debris slides, avalanches, earthflows, mudflows, earth slides and slumps. These movements most commonly occur during times of heavy precipitation. Landslides are the primary mechanism whereby areas of high relief, such as mountains, erode in geologic time.

Laramide Orogeny A time of deformation in the Eastern Rocky Mountains of the United States, whose several phases extended from late Cretaceous into the Tertiary. The orogeny included mountain building, *volcanic* activity, emplacement of *igneous* batholiths and mineralization. Named after the Laramie Formation of Wyoming and Colorado.

lateral moraines *Moraines* formed along sides of glaciers.

lava Molten *magma* which issues from a *volcanic* vent or fissure. Also, the same material solidified by cooling.

lava flow Lateral, surficial outpouring of molten lava from a vent or fissure. Also the solidified body of rock so formed.

limestone *Rock* composed largely of the *mineral* calcite ($CaCO_3$). Most limestones are *marine* in origin. Limestones may also form in fresh-water lakes.

lithification   Conversion of newly deposited sediment into solid rock by processes such as cementation, compaction and crystallization.

lithosphere   In the *Theory of Plate Tectonics*, the lithosphere is composed of the *crust* <u>and</u> the uppermost rigid layer of the *mantle*, overlying the plastic asthenosphere layer.

lode *Mineral* deposit consisting of a vein or veins in consolidated rock as opposed to *placer* deposits in unconsolidated materials.

mafic Said of an *igneous rock* composed chiefly of one or more dark-colored *ferromagnesian minerals*. Basalt and *gabbro* are mafic rocks.

magma Naturally occurring molten rock generated within the earth. *Igneous rocks* are derived from magma.

magnetite   Iron oxide (Fe_3O_4). It is magnetic and an ore of iron.

mammoth   Extinct elephant widely distributed in the Pleistocene. Distinguished by large size, very long upcurved tusks and well-developed body hair.

mantle   Zone in earth's interior between the *crust* and the core. The mantle is largely composed of *ferromagnesian minerals* forming a rock type called *peridotite*.

marble *Metamorphic rock* composed of the *mineral calcite* or sometimes *dolomite*. The *parent rock* (before *metamorphism*) was a *limestone* or *dolomite*.

marginal marine   Refers in general to coastal areas such as coastlines, bays, reefs, *lagoons*, estuaries, and other shallow *marine* areas.

marine   Of, belonging to, or caused by the sea.

mass extinctions   Sudden extinction, as evidenced by the *fossil* record, of a number of species at a specific point in geologic time.

mass wasting Movement, caused by gravity, in which bedrock, rock debris or soil moves downslope in bulk. Includes movements ranging from very slow to fast, such as creep. earthflows, rock slides, avalanches and rock falls. A general term for gravitational earth movements is landslide. Mass wasting is the primary geologic process whereby areas of high relief are reduced, in geologic time, to low relief. Colorado has many areas where landslides occur.

mastodon Extinct animal related to the elephant. It was abundant in the middle Tertiary.

medial moraines When two glaciers merge, the two inside *lateral moraines* join forming a medial moraine in the middle of the glacier.

mesa *Landform* which is flat-topped or nearly so and bounded on all sides by abrupt or steeply sloping *erosion* scarps. Capped by layers of resistant rocks (usually *lavas*).

metamorphism   The word means "changed form".   If rocks are subjected to sufficient heat and pressure, they may be changed.   Old *minerals* decompose and new ones are created.   In addition to mineral changes, platy minerals may become oriented with their long dimensions parallel (foliation). Minerals may also be reorganized into alternating light and dark layers.

metamorphic   *Rock* which has undergone metamorphism.

metasedimentary   Said of *rocks* of sedimentary origin showing evidence of metamorphism.

methane   Chemical compound composed of carbon and hydrogen (CH_4).

mica   Group of chemically complex silicate *minerals*.   Micas have perfect cleavage and readily split into thin layers or sheets.   The most common micas are biotite (black) and muscovite (silvery).

micaceous   Containing mica.

microtektite   Very small *tektite*.

migmatite   Mixed *igneous* and *metamorphic rock*.   *Gneisses* heated to the point of partial melting create zones of igneous rock interlayered with gneiss.

mineral   Naturally occurring, solid, inorganic substance which has an orderly internal structure and characteristic chemical composition.

mollusk   Solitary shell-bearing invertebrate belonging to the phylum Mollusca. Includes *gastropods*, *pelecypods* and *cephalopods*.

molybdenite   Lead-gray, soft *mineral* composed of *molybdenum* and sulfur (MoS_2). It is the primary ore of molybdenum.

molybdenum   Element used as a steel alloy.

monoclinal folding   Deformation of strata in such a manner as to create monoclines.

monocline   Unit of strata which dips or flexes from the horizontal in one direction only and is not part of an anticline or syncline.   Usually a large feature of gentle dip.

moraine   Mound, ridge or other distinct accumulation of unsorted, unstratified glacial till.   Deposited by glaciers.

mudstone   *Clastic sedimentary rock* composed of *clay* and silt grains.   May contain sand grains.   Not laminated and weathers into blocky fragments.

M.Y.A.   Million years ago.

native gold   Gold found uncombined with other elements and in a nongaseous state in nature.   Most native gold, however, does contain some silver as an impurity.

Nemesis, the Death Star   In the Nemesis Theory, a hypothetical twin star orbiting our Sun every 26 million years.   Its passage would disrupt orbits of *comets* in the *Oort Cloud*, causing them to strike the earth and indeed, all the planets and the Sun.   The concept of Nemesis is not popular with astronomers.

neon   Colorless, inert gaseous element occurring in the atmosphere.

NH^3 Chemical formula of ammonia.

nodule Small, hard, and irregular, rounded or tuberous body of a *mineral* or mineral aggregate. Normally has a warty or knobby surface and no internal structure. Usually exhibits a contrasting composition from and a greater hardness than the enclosing sediment or rock matrix in which it is embedded. Most nodules appear to be secondary structures formed after deposition and during or after lithification. Nodules composed of the mineral chert are common in *limestones* and *dolomites*.

nonmarine Not having anything to do with the sea. Continental.

nuee ardentes Swiftly flowing and very hot gaseous cloud, sometimes incandescent, erupted from a *volcano* and containing ash and other *pyroclastics* in its lower part. This lower part of the nuee ardente is comparable to an ash flow. *Welded tuffs* are sometimes created by nuee ardentes.

obsidian Black or dark-colored *volcanic* glass, usually of *rhyolite* composition characterized by conchoidal fracture.

oceanic trench Oceanic trenches are the deepest areas on the oceanic floors. They are long linear features and parallel oceanic *subduction zones*. Trenches are the surficial expression of subduction zones where one plate is subducting below another.

oil shale *Kerogen*-bearing, finely laminated brown or black rock which will yield liquid or gaseous hydrocarbons upon distillation. The rock has the appearance of a shale, but in fact contains considerable calcite and *dolomite*, as well as *clay*. It could be called a shaley *limestone*, but the name "oil shale" appears to be the name of choice.

olivine Olive-green to .brown *mineral* with the chemical composition $(Mg,Fe)_2SiO_4$. Olivine is a common rock-forming mineral of basic, ultrabasic and low-silica *igneous* rocks, including *gabbro*, basalt, peridotite and dunite. It metamorphoses to *serpentine*. Much of the rocks of the earth's *mantle* contain olivine. The gem variety of olivine is known as peridot.

oolitic limestone Limestone composed of tiny, round balls of calcium carbonate.

Oort Cloud The Oort Cloud is a cloud of many thousands, perhaps millions, of *comets* orbiting within our solar system beyond the planet Pluto.

Ophiomorpha *Trace fossil* consisting of a knobby burrow in sandstones. Believed to have been formed by a shrimp inhabiting beaches and shallow water environments. This fossil is considered to be an indicator of shoreline environments.

osmium Element and a member of the platinum metals group. These metals are very rare in the earth's *crust*.

overthrust fault Low angle *thrust fault* of large scale. It is formed by compressive forces which thrust older rocks up and over younger rocks.

paleo- Combining form denoting the attribute of great age or remoteness with regard to time, or involving ancient conditions, or of ancestral character, or dealing with *fossil* forms.

paleomagnetism Study of natural remnant magnetization in order to determine the intensity and direction of the earth's magnetic field in the geologic past. Magnetic *minerals* in *volcanic lavas* reflect the direction of magnetic north at the time the lava cooled and solidified into rock. Importantly, these studies may be used to determine the approximate migrations of continents as plates moved through geologic time.

paleomudcracks Ancient mudcracks preserved in rocks.

paleoraindrop impression Ancient raindrop impressions preserved on bedding plane surfaces in rock. They are abundant in the *eolian* sand dune facies of the Lyons Formation in Colorado.

paleopalynology Study of *fossil pollen* and spores, and additionally other microscopic *fossil* organic bodies. Very useful for studying fine-grained *rocks* such as *shales*.

palynology Geologists use the word, palynology, when they really mean to say paleopalynology.

Pangaea Ancient supercontinent existing during the Pennsylvanian Period including all continental *crust* of the earth. Pangaea began to break up due to *plate tectonics* during the Permian Period and the continents slowly migrated to their present positions.

panning Technique of prospecting for heavy metals, e.g. gold, by washing *placer* or crushed vein material in a pan. Lighter *rock* and *mineral* particles are washed away, leaving heavy metals behind in the pan.

parent rock Source *rock*. With regard to *metamorphic* rocks, the original rock <u>before</u> *metamorphism* into a new, distinct *rock*.

paternoster lake One of a linear chain or series of small circular lakes occupying rock basins, usually at different levels, in a glacial valley. Lakes are separated by morainal dams or higher areas of rock, but are connected by streams, rapids or waterfalls. They resemble a string of beads.

patterned ground Certain well-defined, more or less symmetrical *landforms*, such as rock circles, polygons, nets and stripes, characteristic of surficial material subject to intensive frost action. Most common in polar regions. In Colorado, common in mountain areas above timberline.

pediment Broad, flat or gently sloping, rock-floored erosional surface. Typically developed by *subaerial* agents (including running water) in an arid or semiarid region at the base of an abrupt and receding mountain front or *plateau* escarpment. Underlain by bedrock.

pegmatite Exceptionally coarse-grained *intrusive igneous rock*. The majority of pegmatites have the chemical composition of *granite* (*feldspar*, *quartz*, *biotite* and *hornblende*).

pelecypod Type of shell-bearing *mollusk*. Similar to brachiopods but with a different shell structure. Evolved in the Paleozoic and became very abundant in the Mesozoic and Cenozoic eras. *Fossil* pelecypods are very common in Mesozoic and Cenozoic rocks. All pelecypods may be called "*clams*".

permafrost Permanently frozen ground. Occurs in polar regions and in high alpine areas. In Colorado, permafrost underlies *talus* on north-facing mountain slopes above timberline.

permanent snowfield In Colorado, an area of snow in mountainous regions above snowline and persisting throughout the year. Snowfields are stationary as opposed to glaciers, which move downhill.

permeable Property or capacity of earth materials to transmit a fluid through pore spaces.

Permo-Penn An abreviation used by geologists for the Permian and Pennsylvanian periods combined.

phlogopite Magnesium *mica*, similar to *biotite*, but containing little iron. It is a product of *metamorphism* and occurs in *crystalline limestone* or *dolomite*. It also is found in *serpentine*.

photosynthesis In green plants, use of light energy (aided by the green pigment chlorophyll) for building new organic matter from water and carbon dioxide.

phyllite *Rock* composed of fine-grained *micas* commonly formed by regional *metamorphism*. Intermediate in *metamorphic* grade between *slate* and *mica schist*. Minute crystals of micas impart a silky sheen to the surfaces of cleavage or schistosity.

piedmont Area, plain, slope, bajada or other feature at the base of a mountain or mountain range. The eastern plains of Colorado are a large piedmont at the foot of the Rocky Mountains.

pinnacle Tall, very slender, tapering or pointed tower or spire-shaped pillar of *rock*, either isolated or at the summit of a mountain or hill.

pitchblende Massive, amorphous or cryptocrystalline variety of the *mineral* uraninite, an ore of uranium.

placer Surficial *mineral* deposit formed by mechanical concentration of mineral particles from weathered debris. A majority of placers are formed in river deposits. Gold, a very heavy metal, is found in placer deposits.

plate tectonics Theory in which the *lithosphere* of the earth is divided into a series of fragments or plates which move. Plates are created along certain boundaries (*spreading ridges*) and destroyed at other boundaries (*subduction zones*). In places, plates slide past each other along *transform faults*.

plateau Broadly, any comparatively flat area of great extent and elevation. Commonly limited on at least one side by an abrupt descent or cliff face.

polished glacial pavement Glaciers pick up and move enormous quantities of rocks and other earth materials. When fine-grained material such as sand, silt and clay is carried in the ice at the base of the glacier, it smoothes and polishes the surface of the bedrock below the glacier. Such polished surfaces on bedrock are often observed in areas that have been glaciated.

porous Said of a sediment or *rocks* containing void spaces (sometimes called pore spaces). Porosity is the percentage of void spaces per unit volume of earth material.

provenance Place of origin: specifically the area from which the constituent materials of a *sedimentary rock* were derived by *erosion* and transport.

pyrite Iron sulfide *mineral* (FeS_2). Pyrite may be an *igneous*, *metamorphic* or *sedimentary mineral*.

pyroclastic Pertaining to *clastic rock* material formed by *volcanic* explosion or aerial expulsion from a volcanic vent. Not a *flow rock*.

pyroxene Group of dark, rock-forming chain silicate minerals with complex chemistry. Occur in mafic igneous rocks and some metamorphic rocks. *Augite* is the most common pyroxene mineral.

quartz monzonite Essentially a *granite* with less potassium *feldspar* (orthoclase) and more sodium feldspar (plagioclase). Also could be described as being intermediate between a granite and a *diorite*.

quartz Silicon dioxide *mineral* (SiO_2). The earth's *crust* is largely composed of quartz and other silica-rich minerals.

quartzite Low-rank *metamorphic rock* whose *parent rock* is sandstone. Sandstone is converted to a quartzite when subjected to high temperatures which weld the grains of *quartz* together, producing a harder rock. Quartzites may be visually identified by the following: fractures in quartzite will pass through the sand grains whereas fractures in a sandstone will go around them.

refractory shale *Siliceous* shale rich in hydrous aluminum silicates. Capable of withstanding high temperatures without deforming. Useful for the manufacture of refractory ceramic products.

regional metamorphism General term for metamorphism affecting an extensive region.

restricted Term used to describe bodies of water separated from the open ocean. Such restricted waters will evaporate, leaving behind deposits of *evaporites*.

rhyolite Fine-grained *extrusive igneous rock* which is the *extrusive* equivalent of *granite*. Rhyolite is high in *silica* content and is associated with explosive *volcanoes*. Rhyolitic *magma* is believed to be derived from melting of silica-rich continental rocks.

ridge-former Said of *rocks* resistant to *erosion* which tend to form high areas, such as ridges where rock strata are inclined from horizontal.

Rio Grande Rift The Rio Grande Rift is a rift where the continent is being separated by a linear zone of up-welling basalt *magma* from the *mantle*. Should this continue for millions of years, the continent could be split in half and the ocean would move in. It would at that time be called a *spreading ridge*. The rift is currently widening a few millimeters per year. In New Mexico, associated basalt *lavas* have flowed out in numerous places several times during the last few thousand years.

ripplemarks Small-scale subparallel ridges and troughs formed in loose sand by wind or water currents or waves. Also, such forms preserved in consolidated *rock*.

roches montonnees Small, elongate, protruding knob or hillock of bedrock (most commonly *granite*) sculpted by a large glacier.

rock glacier Mass of poorly sorted angular boulders and fine material cemented by interstitial ice a meter or so below the surface. Occurs in high mountains in a *permafrost* area. The rock material is derived from a *cirque* wall or other steep cliff by frost action. Although largely rock, it has the general appearance and slow movement of a small *valley glacier*. A series of transverse, arcuate and rounded ridges formed by movement are easily recognizable on the upper surface of the rock glacier.

rock   Aggregate of one or more *minerals*.   Because of this variability, rocks are not as precisely defined as minerals.   Rocks are classified based upon mode of origin, texture and mineral composition.   The three major types of rocks are *igneous*, *metamorphic* and *sedimentary*.

sandstone   A *clastic sedimentary rock* composed of grains of sand size (1/16 millimeter-2 millimeter) set in a matrix of silt or *clay*.   It is cemented by precipitated *quartz* or *clay* and less commonly by iron oxide or calcium carbonate cement.   Grains in most sandstones are composed of fragments of the *mineral* quartz, which is the most stable common mineral.

schist   Strongly foliated *crystalline metamorphic rock* frequently composed of *mica*.

scoria   Vesicular, cindery andesitic or basaltic *lava*.   The vesicular nature of which is due to the escape of *volcanic* gases before solidification.   It is usually heavier, darker, and more *crystalline* than pumice.

sedimentary   *Rock* formed when soft sediment is lithified.   Sediment is composed of rock or *mineral* fragments which have been weathered from previously existing rock.   It is transported by wind, water, ice, or gravity and deposited in unconsolidated layers.   These may be later lithified into hard sedimentary rock.   Sedimentary rocks also include those formed by chemical reactions and by secretions of organisms.

serpentine   Group of *rock*-forming *minerals* having the formula $(Mg,Fe)_3Si_2O_5(OH)_4$.   Serpentines have a greasy or silky luster and a slightly soapy feel.   They are always secondary minerals, derived by alteration of magnesium-rich silicate minerals (especially *olivines*) and are found in both *igneous* and *metamorphic* rocks.

shale   Fine-grained *clastic sedimentary rock* composed of particles smaller than 1/256 millimeter.   Shale is well indurated and finely laminated which gives it fissility, the ability to split into thin layers.   Similar rocks lacking fissility are better called *claystones*.

shellfish   General term for an aquatic invertebrate animal having a hard shell.

shocked quartz   *Quartz* subjected to a high-velocity shock will develop a unique fracture pattern composed of parallel sets of planar fractures, sometimes with multiple sets of different orientation.   High-velocity shocked quartz occurs in crater walls of meteorite impact craters.

silica   Term for oxygen combined with silicon.   The common *mineral quartz* (SiO_2) is pure crystallized silica.

siliceous   Said of a *rock* containing *silica*.   A hard shale containing silt and sand could be called a siliceous shale.

silicic acid   Acid containing silicon (H_4SiO_4).   Formed by what is probably the most common *weathering* reaction on earth - the hydrolysis of *feldspars* by carbonic acid in water.

siltstone   Fine-grained *clastic sedimentary rock* composed of fragments ranging from 1/256 millimeter to 1/16 millimeter in diameter.   Siltstones are not fissile as are *shales* and typically weather into blocky fragments.

slate   Compact, fine-grained, *metamorphic rock* formed from *parent rocks* such as *shale* and *volcanic ash*.   Slate possesses the property of fissility (slaty cleavage) along a plane independent of the original bedding.

slip face Steeply sloping surface on the lee side of a dune, standing at or near the angle of repose of loose sand, and advancing downwind by a succession of slides whenever the angle is exceeded.

sluicing An effective way of mining gold from *placer* deposits. A trough or sluice is constructed with riffles in its bottom. Gold-bearing gravels are shoveled or washed into a sluice. Flowing water carries away the *rock* and *mineral* particles, leaving behind the much heavier gold.

smectite Disapproved name for the montmorillonite group of *clay minerals*. The name, however, is still in use.

stock Small *intrusive igneous* body with an outcrop area of less than 100 square kilometers.

solifluction Slow, viscous, downslope flow of water-logged soil and other unsorted and saturated surface material. Occurs at high elevations in regions underlain by frozen ground (*permafrost*).

sphalerite Brown to black, sometimes yellow or white *mineral* composed of zinc, iron and sulfur [(Zn,Fe)S]. Sphalerite frequently contains other elements as impurities. It is the most common ore of zinc and is often associated with *galena* in mineralized veins.

spinel *Mineral* having the composition of $MgAl_2O_4$. Spinel has great hardness, usually forms octahedral crystals, varies widely in color (from colorless to red, greenish and yellow to black), and is used as a gemstone. It occurs in *metamorphic rocks* and as a constituent in basic *igneous* rocks such as peridotite.

sponge Many-celled aquatic invertebrate belonging to the phylum Porifera. Characterized by an internal skeleton composed most frequently of opaline *silica* and less commonly of calcium carbonate. They range from the Precambrian to the present.

spreading center (or ridge) Boundary or zone where lithospheric plates rift or separate from each other. Molten basalt rises from the *mantle* along these linear boundaries and creates new basaltic plate *rock*. The newly formed rocks then migrate in either direction from the spreading center.

stratigraphically equivalent unit Geologists correlate (with respect to time) *rocks* from one area to rocks (which may be of a different lithology) in another area. For example, shales deposited in an ocean are stratigraphically equivalent to sandstones deposited at the same time on the adjacent land.

stratigraphy Branch of *geology* dealing with the definition and description of major and minor divisions of *rocks* (mainly *sedimentary* but not excluding *igneous* and *metamorphic*) available for study in outcrop. Specifically, the geologic study of the form, arrangement, geographic distribution, chronologic succession, classification and correlation relationships of rocks. It thereby involves interpretation of these features of rock strata in terms of their origin, occurrence, environment of deposition, thickness, lithology, composition, *fossil* content, age, history and paleogeographic conditions.

striations Visible line, scratch or lineation. There are many different types of striations in *geology*. Glacial striations are scratches on bedrock created by moving glaciers. Striations (visible lines) on *mineral* crystal faces are created when twinning occurs (as in plagioclase, one of the *feldspars*). Striations (scratches) occur on *rock* surfaces in *faults* as rocks grind past each other during movement.

subaerial Occurring beneath the atmosphere or in the open air. Not underwater.

subaqueous Under, or in water.

subduction zone Long, narrow belt where lithospheric plates collide. The most dense plate is pushed under the other and descends (subducts) into the *mantle*.

swamp Area of standing water with shrubs and trees. Sediments that accumulate in swamps are generally carbonaceous *clays*, commonly with abundant log, limb and leaf fossils. A shrub and tree root zone should exist in the bottom sediments of a swamp.

talus *Rock* fragments of any size or shape (usually coarse and angular) derived from and lying at the base of a cliff or on a very steep, rocky slope. In Colorado, talus deposits are very abundant on mountain slopes above timberline.

tectonics Study of deformation of *rocks* by tectonic forces. For example, faulting, folding, uplift and mountain building. Tectonic forces are most abundant where lithospheric plates are colliding.

tektite Small, rounded or elongate (some have a curved, teardrop shape), pitted, jet black to olive-greenish or yellowish body of silicate glass of nonvolcanic origin. .Found usually in widely separated areas of the earth's surface and bearing no relation to the local geologic formations. Originally thought to be of extraterrestrial origin, either from the Moon or as meteorites from outer space. They have a chemical composition compatible with the earth's *crust*. Today, the most widely accepted explanation is they were formed by large meteorite impacts. Molten droplets of crustal *rocks* were splashed high into the upper atmosphere where they cooled, solidified and fell to earth.

tellurides Group of minerals composed of tellurium, gold and silver. The most common is sylvanite [$(Au,Ag)Te_2$].

tennantite Blackish lead-gray *mineral* containing copper, arsenic and sulfur ($3Cu_2S.Sb_2S_3$). It sometimes contains iron, zinc, silver or cobalt replacing part of the copper.

terminal moraine An end *moraine*, extending across a glacial valley as an arcuate or crescentic ridge, which marks the farthest advance or maximum extent of a glacier.

terrigeneous Of, or having to do with land or continents. The term is commonly applied to *nonmarine* or *marine* sediments derived from *erosion* of nonmarine land or *rocks*.

tetrahedrite Copper-antimony sulfide with a variable chemical composition which has long frustrated mineralogists. An ore of copper and one mineral variety (freibergite) contains silver.

thrust fault Fault with a dip of 45^0 or less and in which older rocks have been moved up and over younger rocks. Caused by horizontal compression.

214

trace fossil Sedimentary structure composed of a disturbance in sediment made by an organism as it went about its life's activities. Preserved in the *rock* after *lithification*. Common trace fossils are tracks, trails, burrows and tubes.

transform fault 1. Strike-slip fault associated with *spreading ridges*. Such faults transform unequal spreading rates into lateral displacement along the axis of the spreading ridge. 2. Plate boundary that ideally shows pure strike-slip displacement.

tuff Compacted *volcanic* deposit composed of small *pyroclastic* fragments of *rocks* and *minerals*. It may contain up to 50% of nonvolcanic sediment such as sand or *clay*.

U-shaped valleys Valley having a cross-sectional form of the letter U. Such valleys were shaped by glaciers. In contrast, valleys formed by rivers exhibit a V-shape.

ultramafics *Igneous rocks* composed mostly of iron and magnesium-rich *minerals*, as for example, peridotite. Rocks of the *mantle* are believed to consist largely of peridotite.

ultraviolet Form of radiant energy from the Sun. It has a wavelength shorter than visible light and longer than X-rays. Ultraviolet has high energy and is reactive with various elements and chemical compounds. It is injurious to living organisms.

unconformity Break or gap in the geologic record. Basically, an unconformity represents missing *rocks* which in themselves represent missing geologic time. Unconformities may be created by a period of *erosion* or nondeposition followed by a resumption of sediment deposition. Unconformities may also be a break between eroded *igneous* or *metamorphic* rocks and overlying sedimentary strata.

valley glacier Glacier that occupies a valley. Common in areas of alpine *glaciation*.

valley former Said of a *rock* unit, such as *shale*, which is easily erodible and thereby forms valleys.

vanadinite Red, yellow or brown *mineral* of the apatite mineral group $[Pb_5(VO_4)_3Cl]$. It is an ore of the elements vanadium and lead.

volcanic ash *Extrusive igneous* unconsolidated material or *rock* composed of fine *pyroclastic* fragments under 2 millimeters in diameter.

volcanic Of or having to do with *volcanoes*. *Rocks* formed by eruptions of volcanoes are said to be volcanic.

volcanic neck A volcanic neck is a shallow, *intrusive* structure apparently formed from *magma* which solidified within the throat of a *volcano*.

volcanic dike Tabular *igneous* intrusion which cuts across the planar structures of the surrounding *rock*.

volcaniclastic Pertaining to a *clastic rock* containing *volcanic* material in any proportion, and without regard to its origin or environment. A useful term to describe *clastic sedimentary rocks* containing volcanic materials.

volcano Vent in the surface of the earth through which *magma* and associated gases and ash erupt. Also, the form or structure, usually conical, that is produced by the ejected material.

weathering Destructive processes by which *rocks* are changed on exposure to atmospheric agents at or near the earth's surface. Weathering processes are described as mechanical or chemical in nature. Mechanical weathering is the process by which large pieces of rock break down into smaller pieces. Chemical weathering is the process by which unstable *minerals* in rock decompose into new minerals which are stable under the existing conditions. Weak acid (carbonic acid) is the primary agent of chemical weathering.

welded tuff *Pyroclastic rock* indurated by the combined action of the heat retained by particles, the weight of overlying material and hot gases. Welded tuffs are generally very hard and exhibit some layering. They are not considered to be *flow rocks*.

xenocryst Similar to a *xenolith* except it is a single crystal or a *crystalline* grain of a *mineral* incorporated in an *igneous* rock and foreign to that igneous rock. Derived from wall *rock* adjacent to the igneous intrusion.

xenolith *Rock* fragment foreign to the *igneous* mass in which it occurs. It is a fragment of wall rock adjacent to an igneous intrusion which has broken off and been incorporated into the *magma*.

INDEX

A close-up view of a portion of Cliff Palace, the largest Anasazi cliff dwelling in Mesaverde National Park.